# SpringerBriefs in Applied Sciences and Technology

## Computational Intelligence

*Series Editor*

Janusz Kacprzyk

For further volumes:
http://www.springer.com/series/10618

Kailash Jagannath Karande
Sanjay Nilkanth Talbar

# Independent Component Analysis of Edge Information for Face Recognition

 Springer

Kailash Jagannath Karande
Electronics and Telecommunication
  Engineering
SKN Sinhgad College of Engineering
Pandharpur
Maharashtra
India

Sanjay Nilkanth Talbar
Electronics and Telecommunication
  Engineering
Shri GGS Institute of Engineering
  and Technology
Nanded
Maharashtra
India

ISSN 2191-530X      ISSN 2191-5318 (electronic)
ISBN 978-81-322-1511-0      ISBN 978-81-322-1512-7 (eBook)
DOI 10.1007/978-81-322-1512-7
Springer New Delhi Heidelberg New York Dordrecht London

Library of Congress Control Number: 2013940094

Springer is part of Springer Science+Business Media (www.springer.com)

*Dedicated to my beloved daughter Kanishka*
- Kailash J. Karande

# Preface

An edge in an image is a contour across which the brightness of the image changes abruptly. In image processing, an edge is often interpreted as one class of singularities. In a function, singularities can be characterized easily as discontinuities where the gradient approaches infinity. However, image data is discrete, so edges in an image are often defined as the local maxima of the gradient. These definitions become the way to lead the research work in this book. Edge detection is an important task in image processing. It is a main tool in pattern recognition, image segmentation, and scene analysis. An edge detector is basically a high pass filter that can be applied to extract the edge points in an image.

Edge information plays vital role in many applications of image processing area. To the best of our knowledge, there is hardly any reported research work on face recognition using edge information as features for face recognition with ICA algorithms. Here, we have used edge detection as a feature extraction method to extract edges information from facial images. The independent components are extracted from edge information. These independent components are used with classifiers to match the facial images for recognition purpose. In this research work we have explored Canny and LOG edge detectors as standard edge detection methods. Oriented Laplacian of Gaussian (OLOG) method is explored to extract the edge information with different orientations of Laplacian pyramid. Multiscale Wavelet model for edge detection is also proposed and explored in this research work to extract edge information. This book will give future direction for the PG students and researchers in the area of Image Processing and Pattern Recognition.

# Acknowledgments

It is a privilege for me to have been associated with Prof. Sanjay N. Talbar, my guide, during my research work and writing of this book. It is with great pleasure that I express my deep sense of gratitude to him for his valuable guidance, constant encouragement, motivation, support, and patience throughout this research work. His continuous inspiration helped lot for my personal development and shaped my career as a passionate teacher.

I express my gratitude to Prof. M. N. Navale, President, SPSPM for his constant encouragement and strong support during the completion of this book. He is the main source of inspiration for me to broaden my thinking.

I wish to express my deepest sense of gratitude to my parents and all family members for their moral support and blessings, which enabled me to complete this task. My heartful thanks go to Nisha, my wife for her patience and understanding, and to my daughter Kanishka who had to miss many affectionate hours during writing of this book.

Finally, I would like to thank all those who have helped directly or indirectly during the writing of this book.

Kailash J. Karande

# Contents

# Abbreviations and Symbols

$\alpha(x,y)$   Edge Direction
$\sigma$   Standard Deviation
$d(x,y)$   Distance Operator
$g(x,y)$   Local Gradient
Cos   Cosine Distance
FastICA   FastICA Algorithm
FRT   Face Recognition Techniques
ICA   Independent Component Analysis
JADE   Joint Approximate Diagonalization of Eigenmatrices
KICA   Kernel ICA Algorithm
L1   Manhattan Distance
L2   Euclidean Distance
LEM   Line Edge Map
LOG   Laplacian of Gaussian
MAH   Mahalanobis Distance
OLOG   Oriented Laplacian of Gaussian
PCA   Principle Component Analysis

# Chapter 1
# Introduction

## 1.1 Introduction

The automatic recognition of human faces presents a significant challenge to the pattern recognition research community. Typically, human faces are similar in structure with minor differences from person to person. They are actually within one class of human faces. Furthermore, changes in lighting conditions, facial expressions, and pose variations further complicate the face recognition task as one of the difficult problems in pattern analysis. Gao and Leung [1] proposed a novel concept where faces can be recognized using line edge map (LEM). LEM is generated for face coding and recognition. A detailed study on the proposed concept covers all aspects of human face recognition, i.e., challenges involved in face recognition as follows: (a) controlled/ideal condition and size variation, (b) varying lighting condition, (c) varying facial expression, and (d) varying pose. This research demonstrates that LEM, together with the proposed generic line segment [2] distance measure, provides a new way for face coding and recognition. A shape and texture-based enhanced fisher classifier (EFC) for face recognition method was introduced by Liu and Wecher [3]. EFC, a new face coding and recognition method employs the enhanced fisher linear discriminant model (EFM) on integrated shape and texture features by triangular shape edge detection. The two methods, LEM and EFM, work by LEM to extract the information from face, which is similar to all edge detection methods that lead to our motivation for research work presented in this book.

In feature extraction from face images, a set of landmark points are first identified from the human face, which are then used for feature measurement based on area, angle, and distances between them. Extraction of features from the front view may be performed from the edge images [4]. Template matching may be used for extraction of the eyes from the face image, while features such as nose, lips, chin, etc., may be extracted from the horizontal and vertical edge maps of a human face [37]. In feature-based methods, local features such as eyes, nose, distance between them, and lips are segmented [38] which are then used as input

K. J. Karande and S. N. Talbar, *Independent Component Analysis of Edge Information for Face Recognition*, SpringerBriefs in Computational Intelligence, DOI: 10.1007/978-81-322-1512-7_1, © The Author(s) 2014

data for the structural classifier. The extracted edge information from face using edge detection methods provide minute details regarding these local features, which leads to the feature-based approach for face recognition [36].

## 1.2 Face Subspace

Normally, in the face recognition process the first stage is face detection which separates face (skin) area from non-face (non-skin) area. Separation of face and non-face area is the first step in subspace analysis. Subspace analysis for face recognition is based on the fact that a class of patterns of interest, such as the face, resides in a subspace of the input image space. For example, a small image of $64 \times 64$ has 4,096 pixels; these pixels can express a large number of pattern classes such as trees, houses, and faces. Of these many pixels only a few correspond to faces. Therefore, the original image demonstration is highly redundant, and the dimensionality of this illustration could be greatly reduced when only the face patterns are of interest.

The objective of face subspace representation is the first step in face recognition. The face subspace is the feature vector representation in different formats, which is used for comparison or matching purpose in face recognition. The widely used face subspace representation is PCA. In PCA [5] or eigenfaces approach [6], a small number of eigenfaces [7] are derived from a set of training face images using the Karhunen–Loeve (KL) transform or PCA. A face image is powerfully represented as a feature vector of low dimensionality. The facial appearance in such subspace provides more significant and richer information for recognition than the raw image. With minimum size weight vectors or minimum number of eigenvectors, the same information of raw image is produced in a powerful way. The use of subspace modeling techniques has significantly advanced the face recognition technology. PCA bases are generally used to obtain independent component analysis (ICA) bases, and its improved recognition accuracy has been proved by many researchers. The original face image, its principle component eigenimage, and independent component image are shown in Fig. 1.1a, b, and c.

There are various other approaches that represent face subspace besides PCA and ICA. In feature-based approach local facial components such as eyes, nose,

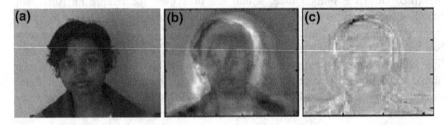

**Fig. 1.1  a** Original face image. **b, c** PC's and IC's image

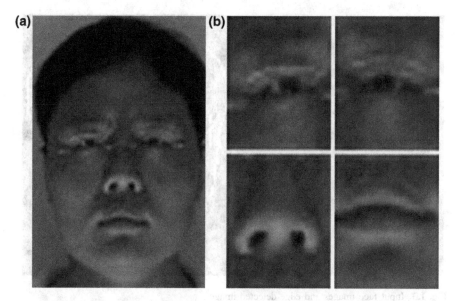

**Fig. 1.2   a, b** Original face image and face subspace with local facial components

lips, and mouth are considered as feature vectors [41]. If we extract these facial components and represent them separately, it becomes yet another representation of face subspace. The original face image and representation with local facial component face subspace is shown in Fig. 1.2a and b.

The edge detected image that contains edge information used for face recognition is another representation of face subspace. Techniques like Canny, Sobel, LOG, Oriented Laplacian of Gaussian (OLOG) [37], and Wavelet-based edge detection [40] are available for edge information extraction from face images. The face image and its edge-detected face image using Canny edge detector, shown in Fig. 1.3, is also face subspace representation used in face recognition.

## 1.3   Edge Detection as Feature Extraction Methods

Edge information plays a vital role in applications in image processing. The edge information is effectively used in iris recognition, fingerprint, texture analysis, and palmistry analysis. To the best of our knowledge, there is hardly any reported research work on face recognition using edge information as features for face recognition with ICA algorithms. Here, we have used edge detection as a feature extraction method to extract edges from facial images. This edge information is further used for extracting the independent components (ICs). The ICs from edge information are used with different classifiers to match the facial images for recognition purposes. Here, different edge detection methods have been used to evaluate the results. Edge detection methods like Canny, Sobel, LOG, Prewitt have been

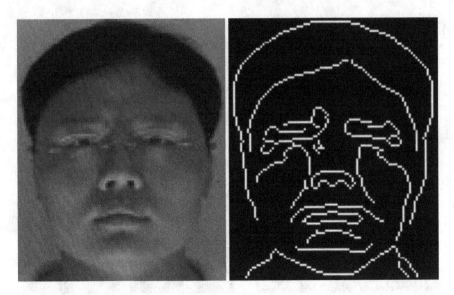

**Fig. 1.3** Input face images and edge detected image

analyzed by researchers. In this study we explore Canny and LOG edge detectors. The OLOG method has been effectively used by researchers [8] for texture analysis with different orientations of laplacian pyramid. Here, we have explored the usefulness of the OLOG technique [37] to extract edge information. The multiscale wavelet model for edge detection proposed by Li [6] is an approach for edge detection explored in this study on facial images to extract edge information.

An edge in an image is a contour across which the brightness of the image changes abruptly. In image processing, an edge is often interpreted as one class of singularities. In a function, singularities can be characterized easily as discontinuities where the gradient approaches infinity. However, image data is discrete, so edges in an image are often defined as the local maxima of the gradient. These definitions lead the way to the research work presented in this book. Edge detection is an important task in image processing. It is the main tool in pattern recognition, image segmentation, and scene analysis. An edge detector is basically a high pass filter that can be applied to extract the edge points in an image. This topic has attracted many researchers and many achievements have been made [10, 11]. Edges are also considered as boundaries between different textures.

## 1.4 Technical Challenges Involved in Face Recognition

Face recognition is of high commercial value due to the involvement of many challenges in its process. The properties of face images change under different conditions, even before or after capturing of images using cameras. The quality of

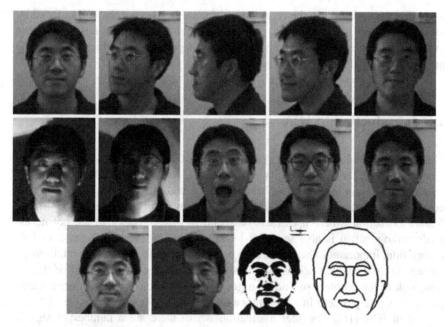

**Fig. 1.4** Intra-subject variations in pose, illumination, expression, occlusion, accessories (e.g. glasses), color, and brightness

face image also depends on the condition under which image is captured as well as on the sensors used to capture the image. The many challenges involved in face recognition are summarized and focused herewith. The major technical challenges in Face Recognition are the topic of research work, and they include large variability in facial appearance with variation in pose and change in Illumination and facial expressions. Whereas shape and reflectance are intrinsic properties of a face object, the appearance of a face is also subject to several other factors, including the facial pose, camera viewpoint, illumination, and facial expression. Figure 1.4 shows an example of significant intra-subject variations caused by these factors. In addition to these, various imaging parameters such as aperture, exposure time, lens aberrations, and sensor spectral response also increase intra-subject variations and affect the quality of face image. Variations between the images of the same face due to illumination and viewing direction are almost always larger than the image variation due to change in face identity [12]. This variability makes it difficult to extract the basic information of face images.

Many approaches to face recognition have been developed that include subspace-based methods such as eigenface [5] and Fisher faces [13], neural network-based recognition [14, 19], morphological elastic graph matching (EGM) [15], linear discriminant analysis (LDA) [16], line-based face recognition [4], and hidden Markov model (HMM) [15]. Besides these methods available, face recognition has become one of the major topics in research areas in image processing and pattern recognition in the recent years due to the various challenges faced by researchers like variation in pose, illumination, and facial expressions. In addition to these challenges are those involving similar images or twins, which becomes the new area and challenge for face recognition.

## 1.5 Introduction to ICA

The new technique of ICA [17, 18] was originally developed to deal with problems that are close to the cocktail-party problem [18]. As an effective approach to the separation of blind signals, ICA has attracted broad attention and has been successfully used in many other fields. One of the promising applications of ICA is feature extraction, where it extracts independent image bases that are not necessarily orthogonal; it is also sensitive to high order statistics. In the task of face recognition, important information may be contained in the high order relationship among pixels. Thus ICA seems a promising face feature extraction method. Let us now look at the problem of finding a good representation, from a different viewpoint. This is a problem in signal processing that also shows the historical background of ICA [18]. Consider a situation where there are a number of signals emitted by some physical objects or sources. These physical sources could be, for example, different brain areas emitting electric signals; people speaking in the same room, thus emitting speech signals; or mobile phones emitting radio waves. Assume further that there are several sensors or receivers. These sensors are in different positions, so that each records a mixture of the original source signals with slightly different weights. For the sake of simplicity of exposition, let us say there are three underlying source signals, and also three observed signals. Denote by $x_1(t)$, $x_2(t)$ and $x_3(t)$ the observed signals, which are the amplitudes of the recorded signals at time point $t$, and by $s_1(t)$, $s_2(t)$ and $s_3(t)$ the original signals. The $x_i(t)$ are then weighted sums of the $s_i(t)$ where the coefficients depend on the distances between the sources and the sensors:

$$
\begin{aligned}
x_1(t) &= a_{11}s_1(t) + a_{12}s_2(t) + a_{13}s_3(t) \\
x_2(t) &= a_{21}s_1(t) + a_{22}s_2(t) + a_{23}s_3(t) \\
x_3(t) &= a_{31}s_1(t) + a_{32}s_2(t) + a_{33}s_3(t).
\end{aligned}
\tag{1.1}
$$

The $a_{ij}$ are constant coefficients that give the mixing weights as shown in Eq. (1.1). They are assumed *unknown*, since we cannot know the values of $a_{ij}$ without knowing all the properties of the physical mixing system, which can be extremely difficult in general. The source signals $s_i$ are *unknown as well*, since the problem is that we cannot record them directly. What we would like to do is to find the original signals from the mixtures $x_1(t)$, $x_2(t)$ and $x_3(t)$. This is the blind source separation (BSS) problem. *Blind* means that we know very little, if anything, about the original sources.

We can safely assume that the mixing coefficients $a_{ij}$ are different enough to make the matrix that they form invertible. Thus, there exists a matrix $\mathbf{W}$ with coefficients $w_{ij}$, such that we can separate the $s_i$ as represented by Eq. (1.2),

$$
\begin{aligned}
s_1(t) &= w_{11}x_1(t) + w_{12}x_2(t) + w_{13}x_3(t) \\
s_2(t) &= w_{21}x_1(t) + w_{22}x_2(t) + w_{23}x_3(t) \\
s_3(t) &= w_{31}x_1(t) + w_{32}x_2(t) + w_{33}x_3(t)
\end{aligned}
\tag{1.2}
$$

Such a matrix could be found as the inverse of the matrix that consists of the mixing coefficients $a_{ij}$ in Eq. (1.1), if we knew those coefficients $a_{ij}$. Now we see that in fact this problem is mathematically similar to the one where we wanted to find a good representation for the random data in $x_i(t)$. Indeed, we could consider each signal $x_i(t), t = 1,........,T$ as a sample of a random variable $x_i$, so that the value of the random variable is given by the amplitudes of that signal at the time points recorded. The question now is: How can we estimate the coefficients $w_{ij}$ in Eq. (1.2)? We want to obtain a general method that works in many different circumstances, and in fact provides one answer to the very general problem that we started with: finding a good representation of multivariate data. Therefore, we use very general statistical properties. All we observe is the signals $x_1$, $x_2$ and $x_3$, and we want to find a matrix $\mathbf{W}$ so that the representation is given by the original source signals $s_1$, $s_2$, and $s_3$.

A surprisingly simple solution to the problem can be found by considering just the statistical independence of the signals. In fact, if the signals are *not gaussian*, it is enough to determine the coefficients $w_{ij}$ so that the signals are statistically independent as shown by Eq. (1.3).

$$
\begin{aligned}
y_1(t) &= w_{11}x_1(t) + w_{12}x_2(t) + w_{13}x_3(t) \\
y_2(t) &= w_{21}x_1(t) + w_{22}x_2(t) + w_{23}x_3(t) \\
y_3(t) &= w_{31}x_1(t) + w_{32}x_2(t) + w_{33}x_3(t).
\end{aligned}
\tag{1.3}
$$

If the signals $y_1$, $y_2$ and $y_3$ are independent, then they are equal to the original signals $s_1$, $s_2$, and $s_3$. Using just this information on the statistical independence, we can in fact estimate the coefficient matrix $\mathbf{W}$. What we obtain are the source signals. We see that from a data set that seemed to be just noise, we were able to estimate the original source signals, using an algorithm that used the information on the independence only. These estimated signals are indeed equal to those that were used in creating the mixtures. Thus, in the source separation problem, the original signals were the "independent components" of the data set.

## 1.6 ICA Algorithms

Here, we have used PCA as baseline preprocessing techniques after edge detection for dimension reduction and whitening of data matrix. In this study we have explored FastICA and Kernel ICA (KICA) algorithms to extract ICs. Here, the research addresses three major challenges faced by face recognition researchers, namely variation in pose, illumination, and facial expressions. After preprocessing and obtaining the eigenvalues and eigenvectors, the next step is to calculate the ICs. A few algorithms used to finds ICs include fixed point ICA algorithm [13], FastICA algorithm [14], KICA algorithm [15], Joint approximate diagonalization of eigenmatrices (JADE) [13], and Blind source extraction-optimization of kurtosis (BSE-K) algorithm [13]. Researchers use FastICA [14] for face recognition as it is faster and gives better results. A few researchers have used KICA [15] and JADE ICA [13] algorithm for face recognition, but the results obtained by these two ICA algorithms are not good enough compared with FastICA. In this study we have explored two ICA algorithms, FastICA and KICA, for extracting ICs and accordingly the comparative results are presented.

## 1.7  FastICA

Here, we used the modified FastICA algorithm [14, 15] for computing ICs. Before computing the ICs we calculate the principle components and then whiten those components to reduce the matrix size. In the previous section we discussed the need to use PCA before applying ICA. After finding the principle components we whiten the eigenvector matrix to reduce the size of matrix and make it square. The main purpose of ICA algorithm is to find $\mathbf{W}$ matrix, which is the inverse of matrix $\mathbf{A}$ as explained by Eqs. (1.1–1.3) in previous section. To estimate several ICs, we need to run the FastICA algorithm several times with weight vectors $\mathbf{w}_1, \ldots, \mathbf{w}_n$. To prevent different vectors from converging to the same maxima we must decorrelate the outputs $\mathbf{w}^T{}_1\mathbf{x}, \ldots, \mathbf{w}^T{}_n\mathbf{x}$ after every iteration.

A simple way of achieving decorrelation is a deflation scheme based on a Gram-Schmidt-like decorrelation [9]. This means that we estimate the ICs one by one. When we have estimated $p$ ICs, or $p$ vectors $\mathbf{w}_1, \ldots, \mathbf{w}_p$, we run the one-unit fixed point algorithm for $\mathbf{w}_{p+1}$, and after every iteration step subtract from $\mathbf{w}_{p+1}$ the projections $\mathbf{w}^T{}_{p+1}\mathbf{w}_j\mathbf{w}_j$, $j = 1, \ldots, p$ of the previously estimated $p$ vectors, and then renormalize $\mathbf{w}_{p+1}$:

$$\text{Let }\ \mathbf{W}_{p+1} = \mathbf{W}_{p+1} - \sum_{j=1}^{p} \mathbf{W}^T{}_{p+1}\mathbf{W}_j\mathbf{W}_j \tag{1.4}$$

and

$$\text{Let }\ \mathbf{W}_{p+1} = \frac{\mathbf{W}_{p+1}}{\sqrt{\mathbf{W}^T{}_{p+1}\mathbf{W}_{p+1}}}. \tag{1.5}$$

In certain applications, however, it may be appropriate to use a symmetric decorrelation, in which no vectors are privileged over others. This can be accomplished by, e.g., the classical method involving matrix square roots

**Table 1.1** Values of independent components (IC) obtained by FastICA

Independent components FastICA

| IC1 | IC2 | IC3 | IC4 | IC5 |
|---|---|---|---|---|
| 4.1941 | 2.2864 | −2.9268 | 0.37954 | 0.20629 |
| 0.85494 | 0.71265 | 0.65535 | 0.6416 | 0.1764 |
| 0.23931 | 0.3667 | 0.93954 | 0.047568 | 0.089064 |
| 0.11136 | −1.313 | −0.35123 | 0.36524 | 0.060164 |
| 0.11136 | −1.313 | −0.35123 | 0.36524 | 0.060164 |
| 0.17897 | 0.41938 | −1.3935 | −1.535 | 0.50269 |
| 0.53329 | −1.0305 | −0.96074 | 3.3739 | 0.078774 |
| 0.39357 | 0.75599 | 1.317 | −6.0059 | 1.5697 |
| 0.77623 | 1.1688 | 0.44573 | 3.284 | 0.58792 |
| 0.77452 | 0.82086 | −4.7952 | 1.1549 | 0.54424 |

$$\text{Let } \mathbf{W} = \left(\mathbf{W}\mathbf{W}^T\right)^{-1/2}\mathbf{W} \tag{1.6}$$

where $\mathbf{W}$ is the matrix $(\mathbf{w}_1, \ldots, \mathbf{w}_n)^T$ of the vectors, and the inverse square root $\left(\mathbf{W}\mathbf{W}^T\right)^{-1/2}$ is obtained from the eigenvalues decomposition of $\mathbf{W}\mathbf{W}^T = \mathbf{FDF}^T$ as $\left(\mathbf{W}\mathbf{W}^T\right)^{-1/2} = \mathbf{FD}^{-1/2}\mathbf{F}^T$. A simpler alternative is the following iterative algorithm:

$$\mathbf{W} = \frac{\mathbf{W}}{\sqrt{\|\mathbf{W}\mathbf{W}^T\|}}. \tag{1.7}$$

As we are using the ICA algorithm the training time required is slightly more compared to other methods. The algorithm developed in this research work required around a few seconds' time as training time. The extracted ICs are given in Table 1.1, where each row represents the first ten values of one independent component. The ICs are represented by images as shown in Fig. 1.5.

## 1.8 Kernel ICA

The second algorithm we explored in this study is the KICA algorithm. For feature extraction, ICA-based linear projection is incompetent to represent the data having nonlinear structure. To address this problem, our idea is to map the data nonlinearly into a feature space, in which the data have a linear structure, i.e., as linearly separable as possible. Then we perform ICA in feature space and make the distribution of data as non-Gaussian as possible. We use kernel tricks to solve the computation of independent projection directions in high-dimensional feature space and ultimately convert the problem of performing ICA in feature space into a problem of implementing ICA in the kernel principal component analysis (KPCA) transformed space. Let us recall the implementation of ICA in observation space.

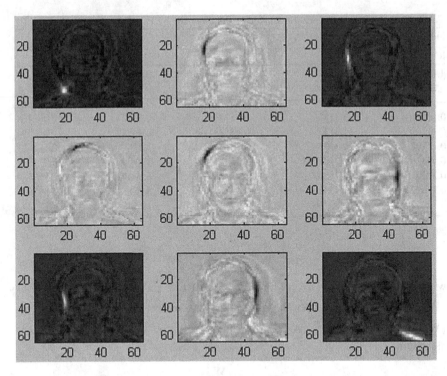

**Fig. 1.5**   A few independent component images obtained by FastICA

Before applying an ICA algorithm on the data, it is useful to do some preprocessing work (e.g. sphering or whitening data). The preprocessing can make the problem of ICA estimation simpler and better conditioned [13]. Generally, a whitened PCA is used to sphere the data and make the transformed data **y** satisfy

$$E\{\mathbf{y}\mathbf{y}^T\} = \mathbf{I}, \quad \text{where } \mathbf{I} \text{ is an identity matrix.} \tag{1.8}$$

Similarly, we can perform PCA in feature space $H$ for data whitening. Note that performing PCA in feature space can be equivalently implemented in input space (observation space) by virtue of kernels, i.e., performing KPCA based on the observation data. Based on this idea, we develop a concise KICA algorithm to implement ICA in feature space.

## 1.9  Sphering of Data Using KPCA

Given an observation sequence $\mathbf{x}_1, \mathbf{x}_2, \ldots, \mathbf{x}_M$ in $R^n$, let us assume their images are centered in feature space, i.e., $\sum_{j=1}^{M} \Phi(x_j) = 0$. The covariance operator on the feature space $H$ can be constructed by

$$S_t^{\Phi} = \frac{1}{M} \sum_{j=1}^{M} \Phi(x_J)\Phi(x_j)^T \tag{1.9}$$

In a finite-dimensional feature space, this operator is generally called covariance matrix. It is not hard to show that $S_t^{\Phi}$ is a positive operator, so nonzero eigenvalues of $S_t^{\Phi}$ are all positive. It is these positive eigenvalues that are of interest to us. Scholkopf et al. [16] suggested a way to find them. Let $Q = [\Phi(x_1), \ldots\ldots, \Phi(x_M)]$ then $S_t^{\Phi}$ can be expressed as $S_t^{\Phi} = (1/M)QQ^T$. Let us form the Gram matrix $\mathbf{R} = \mathbf{Q}^T\mathbf{Q}$, which is an $M \times M$ matrix and its elements can be determined by virtue of the given kernel function $k(\mathbf{x}, \mathbf{y})$, i.e.,

$$\mathbf{R}_{ij} = \Phi(x_i)^T\Phi(x_j) = (\Phi(x_i) \cdot \Phi(x_j)) = k(x_i, x_j). \tag{1.10}$$

Calculate the orthonormal eigenvectors $\gamma1, \gamma2, \ldots\ldots, \gamma m$ of $\mathbf{R}$ corresponding to $m$ largest positive eigenvalues $\lambda1 \geq \lambda2 \geq \ldots\ldots \geq \lambda m$. Then, the $m$ largest positive eigenvalues of $S_t^{\Phi}$ are $\lambda1 \geq \lambda2 \geq \ldots\ldots \geq \lambda m/M$, and the associated orthonormal eigenvectors $\beta1, \beta2, \ldots\ldots, \beta m$ can be expressed as

$$\beta_j = \frac{1}{\sqrt{\lambda_j}}\mathbf{Q}_{yj}, \quad j = 1, \ldots, m. \tag{1.11}$$

Denote $V = (y1, y2, \ldots\ldots ym)$, $\Lambda = \text{diag}(\lambda1, \lambda2, \ldots\ldots, \lambda m)$ and $B = (\beta1, \beta2, \ldots\ldots, \beta m) = QV\Lambda^{-1/2}$ then

$$B^T S_t^{\Phi} B = \text{diag}\left(\frac{\lambda1}{M}, \frac{\lambda2}{M}, \ldots\ldots, \frac{\lambda m}{M}\right) = \frac{1}{M}\Lambda. \tag{1.12}$$

Going one step further, let $P = B\left(\frac{1}{M}\Lambda\right)^{-1/2} = \sqrt{M}QV\Lambda^{-1}$ then

$$P^T S_t^{\Phi} P = I \tag{1.13}$$

Thus, we obtain the whitening matrix $\mathbf{P}$. The mapped data in feature space can be whitened by the following transformation:

$$y = \mathbf{P}^T\Phi(x). \tag{1.14}$$

Specifically,

$$\begin{aligned}
y &= \sqrt{M}\Lambda^{-1}V^T\mathbf{Q}^T\Phi(x) \\
&= \sqrt{M}\Lambda^{-1}V^T[k(x_1, x), k(x_2, x), \ldots\ldots, k(x_M, x)]^T \\
&= \sqrt{M}\Lambda^{-1}V^T\mathbf{R}_x.
\end{aligned} \tag{1.15}$$

Now, let us go back to the problem of centering data. Data centering in input space is easy, but it is difficult to do so in feature space because we cannot explicitly compute the mean of the mapped data in $H$. Fortunately, a slight modification of the above process can achieve this. Denoting an $M \times M$ matrix

$CM = (1/M)_{M \times M}$ and an $M \times 1$ matrix $C1 = (1/M)_{M \times 1}$, let us center the Gram matrix $\mathbf{R}$ and $\mathbf{R}x$ by

$$\tilde{\mathbf{R}} = \mathbf{R} - \mathbf{C}_M \mathbf{R} - \mathbf{R}\mathbf{C}_M + \mathbf{C}_M \mathbf{R}\mathbf{C}_M \tag{1.16}$$

$$\tilde{\mathbf{R}}_x = \mathbf{R}_x - \mathbf{C}_M \mathbf{R}_x - \mathbf{R}\mathbf{C}_1 + \mathbf{C}_M \mathbf{R}\mathbf{C}_1 \tag{1.17}$$

Then, replace $\mathbf{R}$ with $\tilde{\mathbf{R}}$ and get its $m$ largest positive eigenvalues $\lambda 1, \lambda 2, \ldots \ldots, \lambda m$ and the associated orthonormal eigenvectors $\gamma 1, \gamma 2, \ldots \ldots \gamma m$. The whitening transformation is

$$\mathbf{y} = \sqrt{M} \Lambda^{-1} V^T \tilde{\mathbf{R}}_x, \tag{1.18}$$

where

$$V = (y1, y2, \ldots \ldots, ym), \quad \Lambda = \text{diag}(\lambda 1, \lambda 2, \ldots \ldots, \lambda m).$$

## 1.10 Further Processing Using ICA

After whitening, the next task is to find a new unmixing matrix $\mathbf{W}$ in the KPCA-transformed space to recover the independent source $\mathbf{s}$ from $\mathbf{y}$, i.e.,

$$\mathbf{s} = \mathbf{W}\mathbf{y} \tag{1.19}$$

Note that the new unmixing matrix $\mathbf{W}$ should be orthogonal. This can be seen from

$$E\{\mathbf{s}\mathbf{s}^T\} = \mathbf{W}\left[E\{\mathbf{y}\mathbf{y}^T\}\right]\mathbf{W}^T = \mathbf{W}\mathbf{W}^T. \tag{1.20}$$

So far, numerous ICA algorithms have been developed for the calculation of the unmixing matrix. For simplicity, here, we only outline the FastICA algorithm (also called fix-point algorithm) proposed by Hyvärinen et al. [29].

First of all, let us consider the one-unit version of FastICA. The learning rule is to find a maximally "nongaussian" projection direction, i.e., a unit vector $\mathbf{w}$ such that the projection $\mathbf{w}^T\mathbf{y}$ maximizes non-Gaussianity. The non-Gaussianity can be measured via a properly chosen non-quadratic function (contrast function) $G(u)$. Denote the derivative of $G(u)$ by $g(u)$. The basic form of FastICA algorithm is as follows:

Step1 :  Choose anitial (e.g.random) weight vector $\mathbf{w}$;

Step2 :  Let $w_{\text{new}} = E\{\mathbf{y}g(w^T\mathbf{y})\} - E\{g'(w^T\mathbf{y})\}w$;

Step3 :  Let $w = w_{\text{new}}/\|w_{\text{new}}\|$

Step4 :  If not converged, go back to step 2. $\tag{1.21}$

**Table 1.2** Values of independent components (IC) obtained by KICA

Independent components KICA

| IC1 | IC2 | IC3 | IC4 | IC5 |
|---|---|---|---|---|
| 0.8943 | 0.4404 | 0.6818 | 1.8405 | 4.1381 |
| −0.15516 | −0.60953 | 0.62559 | 0.062132 | −0.79509 |
| 0.2152 | 0.9526 | −0.50747 | 0.48524 | 0.29814 |
| −0.14595 | −0.94912 | −0.50747 | 0.77224 | 0.0087162 |
| −0.14595 | −0.94912 | −0.50747 | 0.77224 | 0.0087162 |
| 0.7567 | 3.5962 | 1.0606 | −2.5558 | 0.25575 |
| 1.253 | 0.2503 | 0.8298 | −1.5037 | 0.49797 |
| 2.8795 | 2.6204 | 2.4122 | 0.90275 | −0.5646 |
| 3.7351 | −1.4008 | 1.3957 | 0.31241 | 0.54841 |
| −3.8823 | 0.6116 | −1.7031 | 0.36265 | 0.81557 |

We can estimate a set of orthogonal projection directions $\mathbf{w}_1, \ldots, \mathbf{w}_d$ one by one using the above one-unit FastICA algorithm by virtue of a Gram–Schmidt-like decorrelation scheme. Specifically, after obtaining $k$ directions $\mathbf{w}_1, \ldots, \mathbf{w}_k$, we run the one-unit FastICA algorithm for $\mathbf{w}_{k+1}$, and after every iteration step we normalize $\mathbf{w}_{k+1}$ as

$$w_{k+1} = w_{k+1} - \sum_{j=1}^{k} \left( w_{k+1}^T w_j \right) w_j \quad \text{and}$$

$$w_{k+1} = \frac{w_{k+1}}{\|\mathbf{W}_{k+1}\|} \tag{1.22}$$

The resulting unmixing matrix is $\mathbf{W} = (\mathbf{w}_1, \ldots, \mathbf{w}_d)^T$, $\mathbf{W} = (\mathbf{w}_1, \ldots, \mathbf{w}_d)^T$.

In summary, the ICA transformation in Eq. (1.4) in feature space can be decomposed into two: the whitened KPCA transformation in Eq. (1.17) in input space and the common ICA transformation in Eq. (1.18) in the KPCA whitened space. The first ten values of five independent component vectors are given in Table 1.2, where each row represents ten values of one IC's vector. The ICs are represented by images as shown in Fig. 1.5. The execution time required by KICA algorithm is approximately the same compared with the FastICA algorithm.

If we observe the ICs obtained by FastICA and KICA from Figs. 1.5 and 1.6 there is a difference between them. This gives a clear idea of how the algorithms provide the different information through ICs and how these ICs are spatially different.

## 1.11  Distance Metrics

For face recognition, probe vector is generated by applying the same steps, i.e., preprocessing by PCAs, finding ICs by ICA algorithms, using these ICs and the distance metrics the query image is compared with database images. Mean

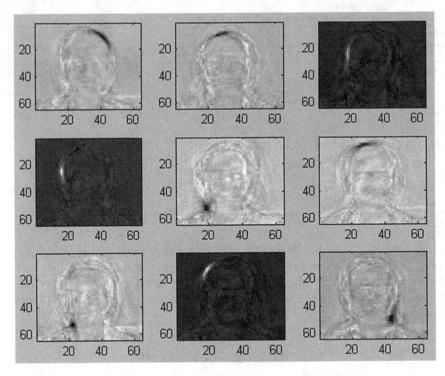

**Fig. 1.6** Independent components' images obtained by KICA

centered probe vector is projected into eigenspace and then IC's space and similarity of test image with each of the training image is checked by nearest neighbor
classifier. Four similarity measures [17–19] used in our study are Manhattan or L1
metric, Euclidean or L2 metric, cosine angle (COS), and Mahalanobis distance
(MAH) given by Eqs. (1.21–1.24) respectively.

$$d_m(x,y) = |x - y| = \sum_{i=1}^{k} |x_i - y_i| \qquad (1.23)$$

$$d_E(x,y) = \|x - y\|^2 = \sum_{i=1}^{k} (x_i - y_i)^2 \qquad (1.24)$$

$$d_{\cos}(x,y) = \frac{x,y}{\|x\|\|y\|} = \frac{\sum_{i=1}^{k} x_i y_i}{\sqrt{\sum_{i=1}^{k} (x_i)^2 \sum_{i=1}^{k} (y_i)^2}} \qquad (1.25)$$

$$d_{\text{Mah}}(x, y) = \sqrt{(x - y)^2 \text{Cov}(x)^{-1}(x - y)} \qquad (1.26)$$

If the distance is small, we say the images are similar and we can decide on the most similar images in the database. The result of face recognition algorithm with all these distance metrics with proper comparison is presented in the research work presented in this book.

## 1.12  Experimental Results

### 1.12.1  Face Database

In order to evaluate how well proposed methods work when dealing with one or a combination of these variation, several face image databases have been built. The most popular face image databases are given in Table 1.3 and details of these databases are given in [20]. The number of people, image size, and number of pictures per person along with other parameters such as the number of conditions are indicative of the complexity of face database and hence also reflects the robustness of face recognition algorithms. This also motivated researchers to generate several face databases that provide as many variations as possible on their images. FERET [21], CMU-PIE [22], AR Faces, Asian face database [23], and Indian face database [24] represent one of the most popular 2-D face image database collections. Each database is designed to address specific challenges covering a wide range of scenarios. For example, FERET represents a good testing framework if one needs large gallery and probe sets, while CMU is indicated more when pose and illumination changes are the main problem finally; AR Faces is the only database providing natural occluded face images; and Asian face database comprises 2-D face images of male and female with illumination, pose, expression, and occlusions. The Indian face database consists of face images with variation in pose and expressions with sufficient number of sample images. In general, face databases of European and American people such as CMU PIE (USA), FERET (USA), AR Face DB (USA), and XM2VTS (UK) have been used for training face recognition algorithms and testing their performance. However, many of the images in databases are not adequately annotated with the exact pose angle, illumination angle, and illuminant color [23]. Also, the faces in these databases have definitely different characteristics from those of Asian and since we wished to check the performance of our algorithm on database of Asian peoples, we used the well-designed Korean face database (KFDB), i.e., Asian and Indian face databases.

Both the databases consist of images of 640 × 480 pixel resolution and 24-bit color depth. They are stored in BMP and JPEG formats. The Asian face database consists of images of 56 male subjects and 51 female subjects with 17 variations: One frontal face image with natural expression; four illumination variations; eight pose variations, and four expression variations. Illumination variations are

**Table 1.3** Most popular face image databases

| Name | Image type RGB/ | Image size | No. of people | Picture or person | Types of conditions | Available | Web address |
|---|---|---|---|---|---|---|---|
| AR Face Database | RGB | 576 × 768 | 126 Male 56 Female | 26 | i, e, o, t | Yes | http://rvl1.ecn.purdue.edu/~aleix/aleix_face_DB.html |
| The yale face database B | Gray scale | 640 × 480 | 10 | 576 | p, i | Yes | http://cvc.yale.edu/projets/alefacesB/yalefacesB.html |
| The yale face database | Gray scale | 320 × 243 | 15 Male 1 Female | 11 | i, e | Yes | http://cvc.yale.edu/projects/yalefaces/yalefaces.html |
| PIE database | RGB | 640 × 486 | 68 | ~608 | p, i, e | Yes | http://www.ri.cmu.edu/projects/project_418.html |
| The UMIST face database | Gray | 220 × 220 | 20 | 19–36 | p | Yes | http://images.ee.umist.ac.uk/danny/database.html |
| FERET | Gray RGB | 256 × 384 | 30.000 | 1,199 | p, i, e, I/O, t | Yes | http://www.itl.nist.gov/iad/humanid/feret/ |
| Asian face database | RGB | 640 × 480 | 300 150 Male 150 Female | 17 | i, p, e, o | Yes | http://nova.postech.ac.Kr/special/imdb/imdb.html |
| Indian face database | RGB | 640 × 480 | 60 30 Males 30 Females | 10 | p, e | Yes | http://www.cs.umass.edu/~vidit/facedatabase |

Image variations are indicated by $i$ illumination, $p$ pose, $e$ expression, $o$ occlusion, $I/O$ indoor/outdoor conditions and $t$ time delay

**Fig. 1.7** Example color images of facial expression variations from Asian face database

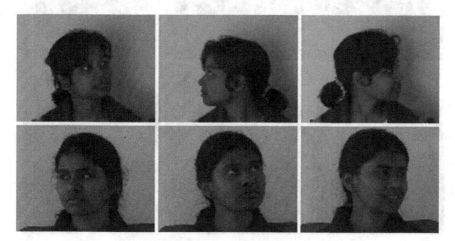

**Fig. 1.8** Example color images of pose variations from Indian face database

obtained by changing lighting directions and illumination color. The lighting direction is changed by using circular arrangement of 8 light sources separated by interval of 45° and illumination color variation is achieved by using fluorescent light and glow light. Pose variations are achieved by capturing the images with seven different cameras.

One camera is placed at the center to capture frontal images and the remaining six cameras are placed with three cameras to the left side of center camera and three cameras to the right side of center camera. The cameras on the left and right sides are separated by interval of 5°, 10°, and 15° with respect to center camera to achieve total variation of 15° on either side of center camera. Four expression variations provided include: Happiness, Anger, Blink, and Surprise with two illumination colors.

The sample images from the database are shown in Figs. 1.7, 1.8, 1.9, and 1.10, respectively, for facial expressions from Asian face database, pose variations from Indian database, illuminations, and pose variations from Asian database. To achieve robust and detailed analysis as mentioned in [23], we report recognition rates for three categories of probe images i.e., illumination variation, pose variation, and expression variation. For pose variation and facial expressions, we used both the databases and for illumination we used only the Asian face database. Recognition accuracy against illumination variations is checked by using one

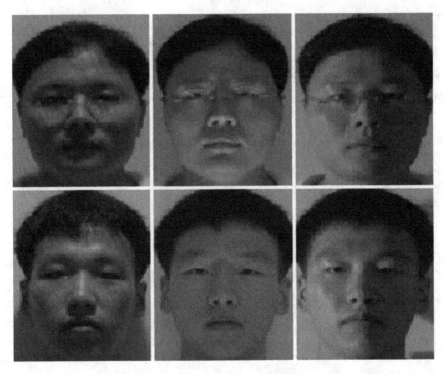

**Fig. 1.9** Example color images of illumination variations from Asian face database

**Fig. 1.10** Example color images of pose variations from Asian face database

frontal face image and two face images with illumination variations per subject for training, i.e., for generation of eigenspace and the remaining two face images with illumination variations per subject are used for testing. Recognition accuracy against pose variations is checked by using one frontal face image and four face images with pose variations per subject for training, i.e., for generation of eigenspace and the remaining four face images with pose variations per subject are used for testing. Recognition accuracy against expression variations is obtained by using one frontal face image and two face images with expression variations per

**Table 1.4** Gallery size and probe set size for three categories

| Size | Probe category size | | |
|---|---|---|---|
| | Illumination variations | Pose variations | Expression variations |
| Gallery size | 168 | 430 | 258 |
| Probe set size | 112 | 344 | 172 |

subject for training, i.e., for generation of eigenspace and the remaining two face images with expression variations per subject are used for testing. The size of the galleries and probe sets for three categories is shown in Table 1.4.

# Chapter 2
# Canny Edge Detection for Face Recognition Using ICA

## 2.1 Introduction

In this chapter we have explored Canny edge detection method to extract edge information. This edge information is further used for obtaining independent components (ICs) using independent component analysis (ICA) algorithms. The ICs are further used as feature vectors for face recognition task.

### 2.1.1 Canny Edge Detection

Canny detector finds edge by looking for local maxima of the gradient of $f(x, y)$. The gradient is calculated using the derivative of a Gaussian filter. The method uses two thresholds to detect strong and weak edges and includes the weak edges at the output only if, they are connected to strong edges. Therefore, this method is more likely to detect true weak edges. The Canny edge detector [25] is the most powerful edge detector provided by function edge. In this method, the image is smoothed using a Gaussian filter with a specified standard deviation, $\sigma$, to reduce noise. The local gradient,

$$g(x, y) = \left[G^2x + G^2y\right]^{1/2} \qquad (2.1)$$

and edge direction,

$$\alpha(x, y) = \tan^{-1}(Gy/Gx) \qquad (2.2)$$

are computed at each point. $Gx$ and $Gy$ computed using Sobel, Prewitt, or Roberts method of edge detection. An edge point is defined to be a point whose strength is locally maximum in the direction of the gradient. Edge also can be defined as discontinuities in image intensity from one pixel to another. The edges for an image are always the important characteristics that offer an indication for a higher

K. J. Karande and S. N. Talbar, *Independent Component Analysis of Edge Information for Face Recognition*, SpringerBriefs in Computational Intelligence, DOI: 10.1007/978-81-322-1512-7_2, © The Author(s) 2014

frequency. Detection of edges for an image may help for image segmentation, data compression, and matching, such as image reconstruction and so on.

Canny edge detector is based on first derivative coupled with noise cleaning. As detection of step edges is influenced by the presence of noise. Therefore, noise smoothing improves the accuracy of edge detection, while adding uncertainty in localizing the edge. Canny edge detector tries to achieve an optimal trade-off between the two thresholds by approximating the first derivative of Gaussian. Canny has considered the following criteria for localizing edges:

(a) There should be low probability of failing to detect a real edge point and, equivalently, low probability of falsely marking non-edge points. That is to maximize the signal-to-noise ratio.
(b) The point marked by the operator as edge points should be as close as possible to the real edge point. That is, minimizing the variance $\sigma^2$ of the zero-crossing position.
(c) The detector should not generate multiple outputs in response to a single edge, i.e., there should be low probability of number of peaks to a given edge response.

Given the good detection, localization, and response to a true edge, the algorithmic steps for Canny edge detection are as follows,

1. Convolve the image $g(r,c)$ with a Gaussian function (select appropriate $\sigma$) to get smooth image $g'(r, c)$, i.e.,

$$\overline{g}(r, c) = g(r, c)^* G(r, c; \sigma) \tag{2.3}$$

2. Apply first difference gradient operator to compute edge strength.

$$d_1 = \frac{1}{2}\{g'(r,c) - g'(r, c - 1) + g'(r - 1, c) - g'(r - 1, c - 1)\} \tag{2.4}$$

$$d_2 = \frac{1}{2}\{g'(r,c) - g'(r - 1, c) + g'(r, c - 1) - g'(r - 1, c - 1)\} \tag{2.5}$$

Then edge magnitude and direction are obtained as before by Eqs. (2.1) and (2.2).
3. Apply non-maximal suppression to the gradient magnitude. This is achieved by suppressing the edge magnitudes not in the direction of the gradient. In fact in Canny's approach, the edge direction is reduced to any one of the four directions. To perform this task for a given point, its gradient is compared with that of points of its $3 \times 3$ neighborhood. If the candidate magnitude is greater than that of neighborhood, the edge strength is maintained, else it is discarded.
4. Apply threshold to the non-maxima suppressed image. Similar to any other edge detection process, the edge magnitudes below a certain value are discarded. However, the Canny's approach employs a clever double thresholding, commonly referred to as hysteresis. In these processes two thresholds, upper and lower are set by the user, so that for a given edgel chain if the magnitude of one edgel of the chain is greater than the upper threshold, all edgels above the

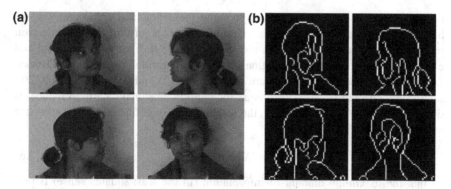

**Fig. 2.1 a** The input face images. **b** Edge detection using Canny operator

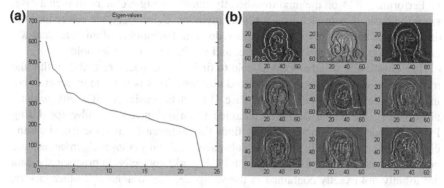

**Fig. 2.2 a** First 25 eigenvalues. **b** First nine eigenimages

lower thresholds are selected as edge points. Canny has not provided any basis for selecting upper and lower thresholds and similar to many such applications, selections of the thresholds are application dependent.

The four sample face images are shown in Fig. 2.1a and the edge detected images using Canny operator are presented in Fig. 2.2b. If we observe the results of Canny edge detector, the orientation of face images are easily recognized and this will help to evaluate the results for pose variation in face recognition.

## 2.1.2 Preprocessing by PCA

There are several approaches for the estimation of the ICA model [35]. In particular, several algorithms were proposed for the estimation of the basic version of the ICA model, which has a square mixing matrix and no noise. Practically, when applying the ICA algorithms to real data, some practical considerations arise and need to be taken into account. To overcome these practical considerations, we have implemented a preprocessing technique in this algorithm that is dimension reduction by principal component analysis (PCA) [39]. That may be useful and even necessary before the application of the ICA algorithms in practice. Overall, face recognition benefits from feature selection of PCA and ICA combination [26].

A common preprocessing technique for multidimensional data is to reduce its dimension by PCA [27]. Basically, the data are projected linearly onto a subspace as given by Eq. (3.1),

$$\tilde{X} = E_n x \qquad\qquad (2.6)$$

so that the maximum amount of information (in the least-squares sense) is preserved. Reducing dimension in this way has several benefits. First, let us consider the case where the number of ICs $n$ is smaller than the number of mixtures; say $m$. Performing ICA on the mixtures directly can cause big problems in such a case, since the basic ICA model does not hold anymore. Using PCA, we can reduce the dimension of the data to $n$. After such a reduction, the number of mixtures and ICs are equal, the mixing matrix is square, and the basic ICA model holds.

The question is whether PCA is able to find the subspace correctly, so that the $n$ ICs can be estimated from the reduced mixtures. This is not true in general, but in a special case it turns out to be the case. If the data consist of n ICs only, with no noise added, the whole data are contained in an $n$-dimensional subspace. Using PCA for dimension reduction clearly finds this $n$-dimensional subspace, since the eigenvalues corresponding to that subspace, and only those eigenvalues, are nonzero. Thus, reducing dimension with PCA works correctly. In practice, the data are usually not exactly contained in the subspace, due to noise and other factors, but if the noise level is low, PCA still finds approximately the right subspace. In the general case, some weak ICs may be lost in the dimension reduction process, but PCA may still be a good idea for optimal estimation of the strong ICs.

Performing first PCA and then ICA has an interesting interpretation in terms of factor analysis. In factor analysis, it is conventional that after finding the factor subspace, the actual basis vectors for that subspace are determined by some criteria, that make the mixing matrix as simple as possible. This is called factor rotation. Now, ICA can be interpreted as one method for determining this factor rotation, based on higher order statistics instead of the structure of the mixing matrix. A well-known benefit of reducing the dimension of the data is that it reduces noise. Often, the dimensions that have been omitted consist mainly of noise. This is especially true in the case where the numbers of ICs are smaller than the number of mixtures. After preprocessing we obtain the eigenvalues and eigenvectors.

Practically, when applying the ICA algorithms to real data, some practical considerations arise and need to be taken into account. To overcome these practical considerations, we have implemented a preprocessing technique in this algorithm i.e., dimension reduction by PCA. That may be useful and even necessary before the application of the ICA algorithms in practice. Overall, face recognition benefits from feature selection of PCA and ICA combination [18]. The few sample face images from Indian face database are given in Fig. 2.1 were used to find out eigenimages and eigenvalues. Using the covariance matrix, the obtained few eigenvalues are represented in Table 2.1. These values are in descending order and approaches to zero represented row wise in the table. This principle of eigenvalue is used to reduce the dimension of eigenvector matrix, and it becomes the advantage of PCA for dimension reduction. The eigenvalues from the Table 2.1 are represented graphically in Fig. 2.2a.

The eigenvector component's values are shown in Table 2.2 and the eigen-images are represented in Fig. 2.2b. Few values of one eigenvector is represented by one row. These eigenimages are nothing but eigenvectors obtained from covariance matrix of input data. The eigenvector matrix is further used by ICA algorithm to extract ICs.

## 2.2 ICA Algorithms

After preprocessing and getting the eigenvalues and eigenvectors next step is to calculate the ICs. Few algorithms used to find ICs are, fixed point ICA algorithm [28], FastICA algorithm [29], Kernel ICA (KICA) algorithm [30], Joint approxi-mate diagonalization of eigenmatrices (JADE) [17], and Blind source extraction–optimization of kurtosis (BSE-K) algorithm [28]. Here, we have used two ICA algorithms for extracting ICs and accordingly the comparative results are pre-sented. These algorithms are FastICA and KICA algorithms. The extracted ICs are represented in Table 2.3, where each row represents few values of one IC vector by FastICA algorithm. The ICs are represented by images as shown in Fig. 2.3a. The ICs are also extracted by using KICA algorithm and are presented in Table 2.4; the ICs are represented by images as shown in Fig. 2.3b.

If we observe the ICs obtained by FastICA and KICA from Fig. 2.3a and b there is difference between them. This gives a clear idea about how these algo-rithms provide the different information through ICs and how these ICs are spa-tially different. The ICs also contain higher order information.

## 2.3 Results with Canny Edge Detector

In the first part, Canny edge detector is used to extract edge information. This edge information is further used by PCA for dimension reduction and whitening data matrix. After whitening ICA algorithms are used to extract ICs further used as feature vectors for face recognition task. The face images are used with variation of pose, illumination, and facial expressions and accordingly results are presented. Here the facial position looking left, right, up, and down are considered. In this experiment, we considered the images from Indian and Asian face database only as

**Table 2.1** Few eigenvalues obtained (row wise)

| Value 1 | Value 2 | Value 3 | Value 4 | Value 5 | Value 6 | Value 7 | Value 8 | Value 9 | Value 10 |
|---------|---------|---------|---------|---------|---------|---------|---------|---------|----------|
| 600.13 | 471.819 | 439.483 | 354.715 | 348.365 | 319.530 | 308.026 | 301.368 | 282.982 | 270.814 |
| Value 11 | Value 12 | Value 13 | Value 14 | Value 15 | Value 16 | Value 17 | Value 18 | Value 19 | Value 20 |
| 264.083 | 253.969 | 246.700 | 240.844 | 239.910 | 227.106 | 214.9116 | 208.8193 | 202.0194 | 194.013 |
| Value 21 | Value 22 | Value 23 | Value 24 | Value 25 | Value 26 | Value 27 | Value 28 | Value 29 | Value 30 |
| 179.3962 | 166.6601 | 0000.0 | 0000.0 | 0000.0 | 00000 | 00000 | 00000 | 00000 | 00000 |

**Table 2.2** First five eigenvector (EV) component values

| | | | | | | | | | | |
|---|---|---|---|---|---|---|---|---|---|---|
| EV1 | -0.0407 | 0.08986 | -0.0814 | 0.10025 | -0.065 | 0.1517 | -0.1395 | 0.2139 | -0.2008 | 0.2491 |
| EV2 | -0.2224 | -0.61629 | -0.0732 | 0.01769 | -0.077 | 0.0103 | 0.0548 | -0.1073 | -0.0015 | -0.0145 |
| EV3 | 0.0296 | 0.05287 | -0.0774 | 0.11022 | 0.1298 | -0.113 | 0.1372 | 0.0378 | -0.0773 | 0.0967 |
| EV4 | 0.6352 | -0.1630 | 0.0106 | -0.0798 | -0.096 | -0.011 | -0.0402 | 0.0316 | -0.069 | -0.0028 |
| EV5 | 0.6352 | -0.16309 | 0.0106 | -0.07986 | -0.096 | -0.011 | -0.0402 | 0.0316 | -0.0693 | -0.0028 |

**Table 2.3** Values of independent components (IC) obtained by FastICA

| IC1 | 0.05436 | −0.00053 | 0.09979 | 0.02951 | 0.02951 | 0.03125 | 0.2304 | 2.8306 | −3.3923 | 0.004232 |
|------|---------|----------|---------|---------|---------|---------|--------|--------|---------|----------|
| IC2 | −0.03703 | 0.071943 | −3.7787 | 0.078979 | 0.078979 | 0.14116 | 0.46677 | 0.39046 | 0.29018 | 0.24986 |
| IC3 | 0.074188 | −0.04077 | 0.11804 | 0.075727 | 0.075727 | 0.21162 | 0.18139 | 0.14876 | 0.0092197 | 0.13955 |
| IC4 | 0.19403 | −0.03045 | 0.067691 | 0.10088 | 0.10088 | 0.11045 | −3.4994 | 1.334 | 0.97321 | 0.072289 |
| IC5 | 0.1246 | −0.10728 | −0.1446 | 0.044354 | 0.044354 | 0.37924 | 0.75166 | −1.9657 | −1.5796 | 0.090231 |

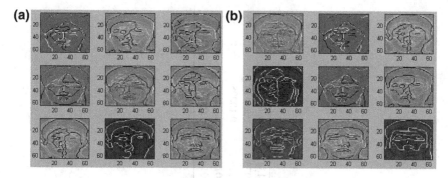

**Fig. 2.3  a** Few ICs images obtained by FastICA and **b** by KICA

the database has sufficient number of sample images for pose variations. Face images for illuminations and facial expressions variation are used from Asian face database. Principle component analysis is one of the old and traditional methods of face recognition, but we have done experimentation with PCA and evaluated results on the same database for comparison purpose only. The test image and the recognized image from gallery set using PCA algorithm is shown in Fig. 2.4a and b.

In these experimentations, we have used two distance metrics L1 and L2. The results of PCA algorithm with L1 and L2 distance metrics with variation in pose, illumination, and expressions are shown in Table 2.5. In this experimentation, we have used four sets of principle components like 25, 50, 100, and 200. The results achieved by PCA method are varying from 60 to 80 %. Specifically, the results achieved by combination of PCA+L2 norms are almost 80 % when the numbers of principle components are 25 for all facial conditions. To the best of our knowledge, there is hardly any reported research work on face recognition using edge information as features for face recognition with ICA algorithms. The recognition accuracy achieved is maximum 80 %, because we calculate the ICs from edge-detected face images, where we lost global information of facial images.

The comparative analysis of two distance metrics is shown in Table 2.5, and same is represented in Fig. 2.5.

Next experimentation for face recognition with variation in pose, illumination, and facial expressions is performed with FastICA algorithm. In this analysis, we have used face images from Indian and Asian face databases with sufficient number of sample face images for pose, illumination, and expression variation. Here, we have used PCA as baseline for FastICA algorithm, where dimension reduction and whitening is done by PCA. The test image and the recognized image from gallery set using FastICA algorithm are shown in Fig. 2.6a and b.

The results of FastICA algorithm with L1 and L2 distance metrics with variation in pose are shown in Table 2.6. In this experimentation, we have used four sets of IC's like 25, 50, 100, and 200. The results obtained by FastICA method are varying from 62 to 84 %. Specifically, the results achieved by combination of FastICA+L2 norms are 84 % under variation of illumination and facial

Table 2.4 Values of independent components (IC) obtained by KICA

| | | | | | | | | | | |
|---|---|---|---|---|---|---|---|---|---|---|
| IC1 | 0.043074 | −0.03921 | 0.07590 | −0.05860 | 0.22064 | −0.0148 | −0.1549 | −0.1341 | −0.21517 | −0.1797 |
| IC2 | −0.005485 | 0.036355 | −0.0034 | 0.045448 | 0.011209 | −0.0138 | −0.00531 | −0.0758 | −0.014747 | −0.0103 |
| IC3 | −0.005711 | −0.08402 | 0.20009 | −0.8345 | 0.046758 | 0.21808 | −0.00771 | 0.22645 | 0.10006 | 0.10538 |
| IC4 | 0.0021956 | −0.02360 | −0.0166 | −0.02161 | −0.01622 | −0.0223 | 0.014882 | −0.0167 | 0.0050791 | −0.0066 |
| IC5 | 0.0021956 | −0.02360 | −0.0166 | −0.02161 | −0.01622 | −0.0223 | 0.014882 | −0.0167 | 0.0050791 | −0.0066 |

**Table 2.5** Result with PCA + Canny

| No of principle components | Pose variation | | Illumination change | | Facial expression | |
|---|---|---|---|---|---|---|
| | L1 | L2 | L1 | L2 | L1 | L2 |
| 25 | 76 | 80 | 76 | 80 | 80 | 80 |
| 50 | 70 | 72 | 68 | 70 | 72 | 68 |
| 100 | 65 | 67 | 64 | 67 | 66 | 65 |
| 200 | 61 | 61.5 | 59.5 | 62 | 62 | 63 |

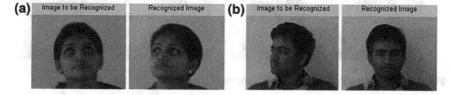

**Fig. 2.4  a–b** Input and output image of algorithm

**Table 2.6** Result with FastICA + Canny

| No of independent components | Pose variation | | Illumination change | | Facial expression | |
|---|---|---|---|---|---|---|
| | L1 | L2 | L1 | L2 | L1 | L2 |
| 25 | 76 | 80 | 80 | 84 | 76 | 84 |
| 50 | 72 | 74 | 72 | 76 | 70 | 76 |
| 100 | 67 | 70 | 66 | 70 | 66 | 72 |
| 200 | 62.5 | 63 | 63.5 | 64 | 63 | 64.5 |

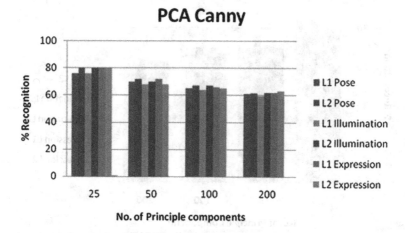

**Fig. 2.5** Percentage recognition accuracy with PCA + Canny

expressions when the numbers of principle components are used 25. The combination of edge information and ICA is a novel idea and as per our study there are no existing results available for comparison. The comparative analysis of two distance metrics is shown in Table 2.7, and same is represented in Fig. 2.7. In this experimentations, we have explored one more ICA algorithm i.e., KICA algorithm

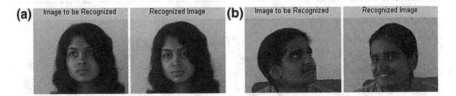

**Fig. 2.6  a–b** Input and output image of algorithm

**Table 2.7**  Result with KICA + Canny

| No of independent components | Pose variation | | Illumination change | | Facial expression | |
|---|---|---|---|---|---|---|
| | L1 | L2 | L1 | L2 | L1 | L2 |
| 25 | 68 | 72 | 72 | 76 | 76 | 80 |
| 50 | 64 | 66 | 64 | 68 | 68 | 70 |
| 100 | 62 | 63 | 61 | 63 | 64 | 65 |
| 200 | 57.5 | 59 | 59 | 60 | 61.5 | 62.5 |

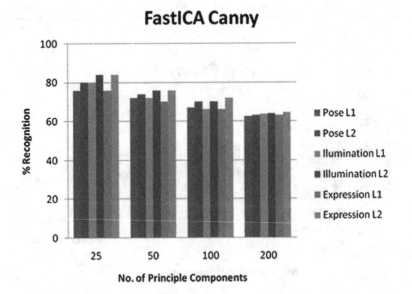

**Fig. 2.7**  Percentage recognition accuracy with FastICA + Canny

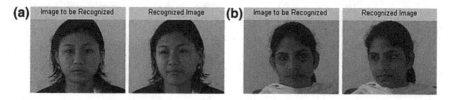

**Fig. 2.8  a–b** Input and output image of algorithm

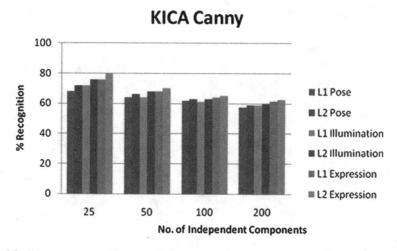

**Fig. 2.9** Percentage recognition accuracy by KICA + Canny

for face recognition under variation of pose, illumination, and facial expressions. For validation of the results, we have used face images from Indian and Asian face databases. The number of ICs varies from 25 to 200 for the experimentation done as shown in Table 2.7.

In this study, we have used KPCA as baseline for ICA algorithm. After dimension reduction by KPCA, FastICA algorithm is used to extract ICs. The combination of KPCA + FastICA is referred as KICA. The test image and the recognized image from gallery set using KICA algorithm are shown in Fig. 2.8a and b.

It is observed from Tables 2.6 and 2.7, as the number of ICs increases the results are also affected. The number of false alarm with respect to correct face recognition increases, this effect is due to similarity of face images in database. For similar face images from gallery set algorithm provides same ICs. From the literature survey, we have observed that combination of edge information and ICA algorithm is not used by researchers. This is the new idea for research with combination of edge information and ICA algorithms. The results of our experiment varies from 59 to 80 % and the comparatively it is presented in Table 2.7, and same is represented in Fig. 2.9.

# Chapter 3
# Laplacian of Gaussian Edge Detection for Face Recognition Using ICA

## 3.1 Introduction

This chapter focuses on another approach of edge detection method used for face recognition with ICA algorithms. In this section, technique used for edge detection is Laplacian of Gaussian (LOG). The edge can be sharpened or enhanced by taking second derivative of image intensity. Edge detection of second derivative operator corresponds to the detection of zero crossing. The widely used second derivative operator is the Laplacian edge detector. In one of its useful variations, Laplacian is preceded by the noise smoothing operation commonly known as LOG.

### 3.1.1 Laplacian Edge Detector

Given an image matrix, the Laplacian of the image function is the second order partial derivatives along $x$ and $y$ directions.

$$\nabla^2 g = \frac{\partial^2 g}{\partial x^2} + \frac{\partial^2 g}{\partial y^2} \tag{3.1}$$

In the digital approximation of the second order, partial derivative in $x$ direction is

$$\frac{\partial^2 g}{\partial x^2} = \frac{\partial \{g(r,c) - g(r,c-1)\}}{\partial x} \tag{3.2}$$

$$= \frac{\partial g(r,c)}{\partial x} - \frac{\partial g(r,c-1)}{\partial x} \tag{3.3}$$

$$= g(r,c) - g(r,c-1) - g(r,c-1) + g(r,c-2) \tag{3.4}$$

$$= g(r,c) - 2g(r,c-1) + g(r,c-2). \tag{3.5}$$

K. J. Karande and S. N. Talbar, *Independent Component Analysis of Edge Information for Face Recognition*, SpringerBriefs in Computational Intelligence, DOI: 10.1007/978-81-322-1512-7_3, © The Author(s) 2014

Similarly, the second-order derivative along $y$ direction is given by,

$$\frac{\partial^2 g}{\partial y^2} = g(r,c) - 2g(r-1,c) + g(r-2,c) \tag{3.6}$$

Both the approximating Eqs. (3.5) and (3.6) are centered around the point $(r-1, c-1)$. Conveniently, this pixel could be replaced by $(r, c)$. The corresponding Laplacian masks for convolution are given by

| 0 | 1 | 0 |
|---|----|---|
| 1 | −4 | 1 |
| 0 | 1 | 0 |

It may be necessary to add higher weights at the center pixel. The restriction is that the sum of all the weights of the mask must add to zero.

### 3.1.2 Laplacian of Gaussian Edge Detector

The weight distribution in a Laplacian mask evokes strong response to stray noise pixels. It indicates that some sort of noise cleaning preceded by Laplacian should provide a better result. For noise cleaning, one may employ Gaussian smoothing. Then the resultant algorithm is LOG edge operator.

The algorithm steps are given as follows:

1. Smooth the image intensity $g(r,c)$ by convolving it with a digital mask corresponding to Gaussian function.
2. Apply the Laplacian mask on the smooth image intensity profile.
3. Find the zero crossing in the image subjected to Laplacian second derivative operator.

Mathematically,

$$g''(r,c) = \nabla^2 \{g(r,c) \times G(r,c)\} \tag{3.7}$$

which, following the rule of convolution, becomes

$$g''(r,c) = g(r,c) \times \{\nabla^2 G(r,c)\}. \tag{3.8}$$

Extracting edge points or edgels by detecting zero crossing in the second derivative still suffers from the problem of false alarm i.e., a non-edge pixel may be marked as edge pixel. This is because even a small nonlinear variation in intensity profile gives rise to zero crossing in second-order derivative. The problem may be surmounted by considering both first- and second-order derivatives. In that case, an edgel is said to be present if there exists a zero crossing in second

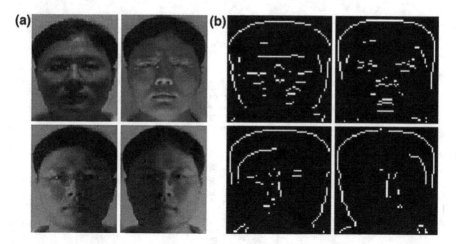

**Fig. 3.1** **a** Input face images. **b** Edge detected images by LOG operator

derivative and if magnitude of first derivative exceeds a thresholds at the same location. The input facial images and edge-detected face images using LOG is presented in Fig. 3.1a and b.

### 3.1.3 Preprocessing by PCA

The preprocessing using PCA before applying the ICA algorithms is well defined and gives better results as discussed in previous chapter. Practically, when applying the ICA algorithms to real data, few practical considerations arise and need to be taken into account. To overcome these practical considerations, we have implemented a preprocessing technique in this algorithm that is dimension reduction by principal component analysis or KL transformation. This is also useful and even necessary before the application of the ICA algorithms in practice. It is proved that face recognition benefits from feature selection using PCA and ICA combination [31]. Using the covariance matrix, the obtained few eigenvalues are represented in Table 3.1. These values are in descending order and approaches to zero, represented row wise in the table. This principle of eigenvalue is used to reduce the dimension of eigenvector matrix, and it becomes the advantage of PCA for dimension reduction. The eigenvalues in the Table 3.1 are represented graphically in Fig. 3.2a.

The eigenvector component's values are shown in Table 3.2 and the eigen-images are represented in Fig. 3.2b. Few values of one eigenvector are represented by one row. These eigenimages are nothing but eigenvectors obtained from covariance matrix of input data. The eigenvector matrix is further used by ICA algorithm to extract independent components (ICs).

**Table 3.1** Few eigenvalues obtained (row wise)

| Value 1 | Value 2 | Value 3 | Value 4 | Value 5 | Value 6 | Value 7 | Value 8 | Value 9 | Value 10 |
|---|---|---|---|---|---|---|---|---|---|
| 456.2398 | 428.2834 | 330.022 | 309.518 | 287.536 | 271.309 | 262.031 | 244.413 | 238.638 | 220.030 |
| Value 11 | Value 12 | Value 13 | Value 14 | Value 15 | Value 16 | Value 17 | Value 18 | Value 19 | Value 20 |
| 209.9756 | 192.9995 | 186.7973 | 175.7689 | 170.416 | 155.2374 | 146.913 | 140.558 | 138.264 | 133.816 |
| Value 21 | Value 22 | Value 23 | Value 24 | Value 25 | Value 26 | Value 27 | Value 28 | Value 29 | Value 30 |
| 126.1620 | 101.2255 | 0000.0 | 0000.0 | 0000.0 | 00000 | 00000 | 00000 | 00000 | 00000 |

**Table 3.2** First five eigenvector (EV) component values

| | | | | | | | | | | |
|---|---|---|---|---|---|---|---|---|---|---|
| EV1 | 9.9637e−016 | 1.6391e−015 | 3.1758e−015 | 8.8618e−016 | 8.8164e−016 | 1.538e−015 | 5.202e−016 | 9.5038e−017 | 1.3018e−015 | 2.247e−017 |
| EV2 | 2.3363e−015 | 1.3795e−015 | 5.2596e−016 | −5.952e−016 | 4.8069e−017 | 7.817e−016 | 3.9327e−016 | 1.2284e−015 | 3.599e−017 | 5.5243e−017 |
| EV3 | 1.0515e−016 | 2.4782e−016 | 5.4046e−016 | 4.3617e−016 | 4.5638e−016 | −8.6741 e−01 | 1.2724e−016 | −3.041e−016 | 4.3993e−016 | 6.2579e−017 |
| EV4 | 4.0719e−015 | 2.0954e−015 | 1.852e−015 | 1.9658e−015 | 1.3572e−015 | 1.6274e−015 | 5.2009e−016 | 2.1671e−015 | 1.1943e−016 | 1.0702e−015 |
| EV5 | 1.0814e−016 | 3.6055e−016 | 1.082e−016 | 6.7227e−016 | 2.6537e−016 | −7.1884 e−01 | 4.5457e−017 | 1.3463e−016 | 5.2345e−016 | 3.2842e−016 |

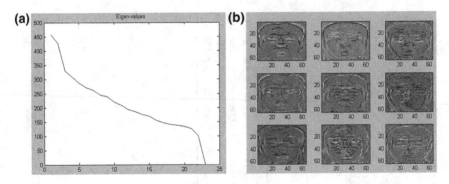

**Fig. 3.2  a** First 25 eigenvalues. **b** First nine eigenimages

## 3.2  ICA Algorithms

After preprocessing and getting the eigenvalues and eigenvectors next step is to calculate the ICs. Few algorithms used to finds ICs are fixed point ICA algorithm [28], FastICA algorithm [29], Kernel ICA (KICA) algorithm [30], Joint approximate diagonalization of eigenmatrices (JADE) [28], and Blind source extraction–optimization of kurtosis (BSE-K) algorithm [28].

Here we have used two ICA algorithms for extracting ICs and accordingly the comparative results are presented. These algorithms are FastICA and KICA algorithms. The extracted ICs are represented in Table 3.3, where each row represents few values of one independent component by FastICA algorithm. The ICs are represented by images as shown in Fig. 3.3a. The ICs are also extracted by using KICA algorithm and are presented in Table 3.4; the ICs are represented by images as shown in Fig. 3.3b.

If we observe the ICs obtained by FastICA and KICA from Fig. 3.4a and b there is difference between them. This gives clear idea about how the algorithms provide the different information through ICs and how these ICs are spatially different.

## 3.3  Results with LOG Edge Detector

In the Chap. 2, the evaluation of face recognition is performed with Canny edge detector to extract edge information. In this part of the experimentation, we explored LOG as edge detection techniques for extraction of edge information. This edge information is further used by PCA for dimension reduction and whitening data matrix. The face images are used with variation of pose, illumination, and facial expressions. Here the facial position looking left, right, up, and down are considered. In this experiment, we considered the images from Indian

**Table 3.3** Values of independent components (IC) obtained by FastICA

| | | | | | | | | | | |
|---|---|---|---|---|---|---|---|---|---|---|
| IC1 | 0.05436 | -0.00053 | 0.09979 | 0.02951 | 0.02951 | 0.0312 | 0.2304 | 2.8306 | -3.3923 | 0.004232 |
| IC2 | -0.03703 | 0.071943 | -3.7787 | 0.078979 | 0.078979 | 0.14116 | 0.46677 | 0.39046 | 0.29018 | 0.24986 |
| IC3 | 0.074188 | -0.04077 | 0.11804 | 0.075727 | 0.075727 | 0.21162 | 0.18139 | 0.14876 | 0.0092197 | 0.13955 |
| IC4 | 0.19403 | -0.03045 | 0.067691 | 0.10088 | 0.10088 | 0.11045 | -3.4994 | 1.334 | 0.97321 | 0.072289 |
| IC5 | 0.1246 | -0.10728 | -0.1446 | 0.044354 | 0.044354 | 0.37924 | 0.75166 | -1.9657 | -1.5796 | 0.090231 |

**Table 3.4** Values of independent components (IC) obtained by KICA

| | | | | | | | | | | |
|---|---|---|---|---|---|---|---|---|---|---|
| IC1 | 0.043074 | −0.03921 | 0.07590 | −0.05860 | 0.22064 | −0.0148 | −0.1549 | −0.1341 | −0.21517 | −0.1797 |
| IC2 | −0.005485 | 0.036355 | −0.0034 | 0.045448 | 0.011209 | −0.0138 | −0.00531 | −0.0758 | −0.014747 | −0.0103 |
| IC3 | −0.005711 | −0.08402 | 0.20009 | −0.8345 | 0.046758 | 0.21808 | −0.00771 | 0.22645 | 0.10006 | 0.10538 |
| IC4 | 0.0021956 | −0.02360 | −0.0166 | −0.02161 | −0.0162 | −0.0223 | 0.014882 | −0.0167 | 0.0050791 | −0.0066 |
| IC5 | 0.0021956 | −0.02360 | −0.0166 | −0.02161 | −0.01622 | −0.0223 | 0.014882 | −0.0167 | 0.0050791 | −0.0066 |

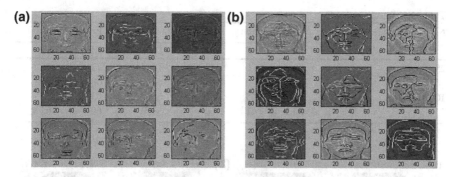

**Fig. 3.3** **a** Few ICs images obtained by FastICA and **b** by KICA

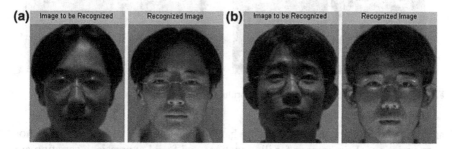

**Fig. 3.4** **a–b** Input and output image of algorithm

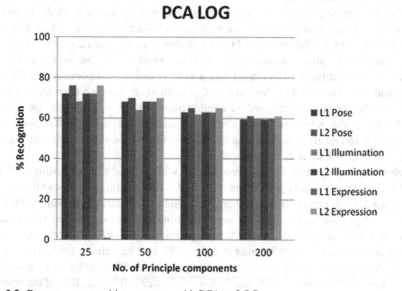

**Fig. 3.5** Percentage recognition accuracy with PCA + LOG

**Table 3.5** Result with PCA + LOG

| No of principle components | Pose variation | | Illumination change | | Facial expression | |
|---|---|---|---|---|---|---|
| | L1 | L2 | L1 | L2 | L1 | L2 |
| 25 | 72 | 76 | 68 | 72 | 72 | 76 |
| 50 | 68 | 70 | 64 | 68 | 68 | 70 |
| 100 | 63 | 65 | 62 | 63 | 63 | 65 |
| 200 | 59.5 | 61 | 60 | 59.5 | 60 | 61 |

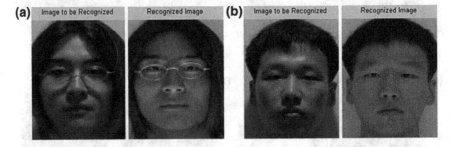

**Fig. 3.6  a–b** Input and output image of algorithm

and Asian face database as the database has sufficient number of sample images for pose variations. Face images for illuminations and facial expressions variation are used from Asian face database. Principle component analysis is one of the old and traditional methods of face recognition, but we have done experimentation with PCA and evaluated results on the same database for comparison purpose only.

The test image and the recognized image from gallery set using PCA algorithm are shown in Fig. 3.5a and b. In these experimentations, we have used two distance metrics L1 and L2. The results of PCA algorithm with L1 and L2 distance metrics with variation in pose, illumination, and expressions are shown in Table 3.5. In this experimentation, we have used four sets of principle components like 25, 50, 100, and 200. The results achieved by PCA method are varying from 60 to 76 %. Specifically, the results achieved by combination of PCA + L2 norms are almost 76 %, when the numbers of principle components are used 25 for pose and expression variations. The comparative analysis of two distance metrics are shown in Table 3.5, and same is represented in Fig. 3.6.

Next experimentation for face recognition with variation in pose, illumination, and facial expressions is performed with FastICA algorithm. In this analysis, we have used face images from Indian and Asian face databases with sufficient number of sample face images for pose, illumination, and expression variation.

In this section, we have used PCA as baseline for FastICA algorithm, where dimension reduction and whitening is done by PCA. The training time required by FastICA algorithm is slightly more as compared with PCA, but it is in few seconds. The combination of PCA and ICA works better is the reflection of the results

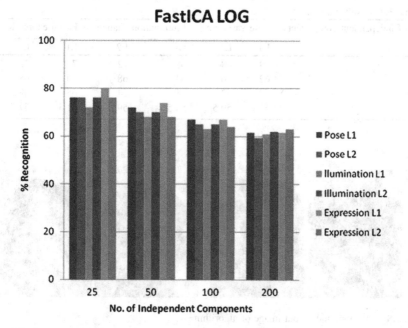

**Fig. 3.7** Percentage recognition accuracy with FastICA + LOG

**Table 3.6** Result with FastICA + LOG

| No of independent components | Pose variation | | Illumination change | | Facial expression | |
|---|---|---|---|---|---|---|
| | L1 | L2 | L1 | L2 | L1 | L2 |
| 25 | 76 | 76 | 72 | 76 | 80 | 76 |
| 50 | 72 | 70 | 68 | 70 | 74 | 68 |
| 100 | 67 | 65 | 63 | 65 | 67 | 64 |
| 200 | 61.5 | 59.5 | 61 | 62 | 61.5 | 63 |

presented in this section. The test image and the recognized image from gallery set using FastICA algorithm are shown in Fig. 3.7a and b.

The results of FastICA algorithm with L1 and L2 distance metrics with variation in pose, illumination, and facial expressions are shown in Table 3.6. In this experimentation, we have used four sets of ICs like 25, 50, 100, and 200. The results obtained by FastICA method are varying from 60 to 80 %. Specifically, the results achieved by combination of FastICA + L2 norms are 80 % under variation of facial expressions, when the numbers of ICs are 25. The combination of edge information and ICA is a new idea and as per our study there is hardly research reported for edge and ICA combination. The comparative analysis of two distance metrics is shown in Table 3.6, and same is represented in Fig. 3.7.

**Table 3.7** Result with KICA + LOG

| No of independent components | Pose variation | | Illumination change | | Facial expression | |
|---|---|---|---|---|---|---|
| | L1 | L2 | L1 | L2 | L1 | L2 |
| 25 | 64 | 68 | 68 | 72 | 76 | 72 |
| 50 | 62 | 66 | 64 | 68 | 70 | 68 |
| 100 | 60 | 62 | 61 | 64 | 65 | 63 |
| 200 | 54 | 59.5 | 54.5 | 59 | 61.5 | 57.5 |

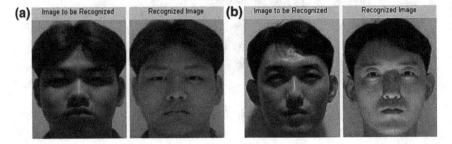

**Fig. 3.8  a–b** Input and output image of algorithm

In this study, we have explored one more ICA algorithm i.e., KICA algorithm for face recognition under variation of pose, illumination, and facial expressions. For validation of the results, we have used face images from Indian and Asian face databases. The number of ICs varies from 25 to 200 for the experimentation done as shown in Table 3.7. In this study, we have used KPCA as baseline for ICA algorithm. The combination of KPCA + FastICA is referred as KICA. The test image and the recognized image from gallery set using KICA algorithm is shown in Fig. 3.8a and b.

It is observed from Tables 3.6 and 3.7, as the number of ICs increases the results also affected. The number of false alarm with respect to correct face recognition increases, this effect is due to similarity of face images in database. For similar face images from gallery set algorithm provides same ICs. From the literature survey and as per our knowledge, the combination of edge information and ICA is new way of research. This is the novel idea for research with combination of edge information and ICA algorithms. The combination of KICA + L1 gives better results. If we compare results of PCA, FastICA, and KICA the results

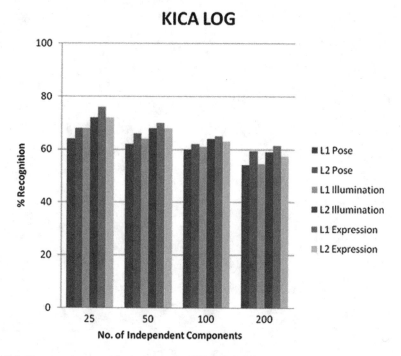

**Fig. 3.9** Percentage recognition accuracy by KICA + LOG

presented by FastICA are better than other two methods. The results of our experiment vary from 54 to 76 % and the comparatively it is presented in Table 3.7, and same is represented in Fig. 3.9.

# Chapter 4
# Oriented Laplacian of Gaussian Edge Detection for Face Recognition Using ICA

## 4.1 Oriented Laplacian of Gaussian

In this chapter, we have explored Oriented Laplacian of Gaussian (OLOG) as edge detection technique. If we use Laplacian pyramid with different orientations, and these oriented pyramids are used as edge information for face recognition is an innovative idea. The concept of multiresolution representation was present in the algorithm called Laplacian pyramid scheme developed by Burt and Adelson [5]. It was developed for the image coding, analysis, reconstruction, and pattern recognition task. The important aim of the decomposition scheme for images is to represent the image efficiently by removing the high correlation existing between neighboring pixels. Various methods have been proposed to achieve this. In the approach, proposed by Burt and Adelson, the predicted value for each pixel is computed as a local weighted average, using a Gaussian-like weighting function from either previously encoded or neighboring pixel. The difference between the original image and low-pass filtered image, called band-pass image, is then encoded. As low-pass image contains only lower spatial frequencies, it can be subsampled by two without loss of information. The resolution change is obtained by the low-pass filter. The scale change is due to subsampling by two, which results the output with one-fourth sample that of original image. The whole process is repeated several times in order to achieve the desired decomposition. The set of low-pass and band-pass components is regrouped in a pyramidal data structure called Gaussian and Laplacian pyramids, respectively. We use the filter subtract decimate (FSD) Laplacian pyramid, which is a variation of the standard Laplacian pyramid. The input image is $G_o(i, j)$, the low-pass filtered images are referred as $G_1(i, j)$ through $G_k(i, j)$ with decreasing resolution and corresponding band-pass versions are referred as $L_1(i, j)$ through $L_k(i, j)$, respectively. An iterative scheme for the creation band-pass composition is given as follow:

$$G_k(i,j) = w(m,n)^* G'_{k-1}(i,j) \qquad (4.1)$$

$$L_k(i,j) = G_{k-1}(i,j) - G'_{k-1}(i,j) \qquad (4.2)$$

where $G'_k(i,j) = G_k(i,j)$ down sample by two as shown in Fig. 4.1.

K. J. Karande and S. N. Talbar, *Independent Component Analysis of Edge Information for Face Recognition*, SpringerBriefs in Computational Intelligence, DOI: 10.1007/978-81-322-1512-7_4, © The Author(s) 2014

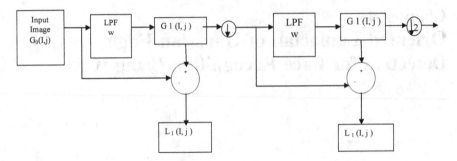

**Fig. 4.1** Gaussian and Laplacian pyramid

The low-pass filter mask $w(m,n)$ is Gaussian in shape and normalized to have the sum of its coefficient equal to 1. We have used one-dimensional five length separable filter with the coefficients {0.0625, 0.25, 0.375, 0.25, and 0.625}. Each higher level image is roughly one-fourth in dimension as its predecessor, due to reduced sample density. Note that image features such as edge and bars enhanced in size: Fine details are prominent in low resolution images, while progressively coarser features are prominent in higher level images.

One of the main advantages of using the pyramid is that the computations are simple and very fast compared to other multichannel filtering. The disadvantage is that, these pyramids do not possess the spatial orientation selectivity in the decomposition process which is very essential in texture analysis. Greenspan et al. [32] overcome the limitation using oriented Laplacian pyramid structure. They obtained oriented Laplacian pyramid with the set of oriented since waves followed by low-pass filtering as defined in [32].

$$O_{n\alpha} = LPF\left\{ \left[ \begin{array}{c} \cos\dfrac{\pi}{2}(x\cos\theta_\alpha + y\sin\theta_\alpha) + \\ j\left(\sin\dfrac{\pi}{2}(x\cos\theta_\alpha + \sin\theta_\alpha)\right) \end{array} \right] L_n(x,y) \right\} \qquad (4.3)$$

where $n$ is the scale coefficient and $\alpha$ represents orientation coefficients and

$$\theta_\alpha = \frac{\pi}{4}\alpha : (\alpha = 0, 1, 2, 3) \qquad (4.4)$$

A three-scale pyramid is used where each of the Laplacian pyramid is multiplied by cosine waves at four orientations (0°, 45°, 90°, and 135°). The orientation and frequency bandwidth of each orientationally tuned band-pass filter is thus 45° and octave, respectively. Figure 4.2 shows the original face images and oriented Laplacian pyramid generated on face image with 45° orientations, respectively. In this study, we have analyzed OLOG with 45° and 135° orientations for extraction of edge information for face recognition with ICA algorithms.

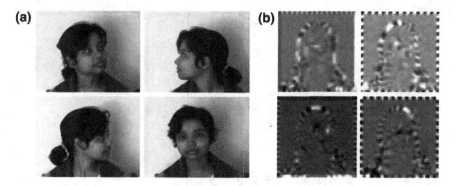

**Fig. 4.2** **a** Input face images. **b** Laplacian pyramid with 45° orientations

## 4.2 Preprocessing by PCA

The preprocessing is briefed in Chap. 2, using PCA. The few sample face images from Indian face database are given in Fig. 4.2a used to find out eigenimages and eigenvalues. Using the covariance matrix, the obtained few eigenvalues are represented in Table 4.1. These values are in descending order and approaches to zero represented row wise in the table. This principle of eigenvalue is used to reduce the dimension of eigenvector matrix, and it becomes the advantage of PCA for dimension reduction. The eigenvalues from the Table 4.1 are represented graphically in Fig. 4.3a.

The eigenvector component values are shown in Table 4.2 and the eigenimages are represented in Fig. 4.3b. Few values of one eigenvector are represented by one row. These eigenimages are nothing but eigenvectors obtained from covariance matrix of input data. The eigenvector matrix is further used by ICA algorithm to extract independent components.

## 4.3 ICA Algorithms

Here we have used two ICA algorithms for extracting independent components and accordingly the comparative results are presented. These algorithms are FastICA and Kernel ICA (KICA) algorithms. The extracted independent components are represented in Table 4.3, where each row represents few values of one independent component by FastICA algorithm. The independent components are represented by images as shown in Fig. 4.3a. The independent components are also extracted by using KICA algorithm and are presented in Table 4.4; the ICs are represented by images as shown in Fig. 4.3b.

If we observe the independent components obtained by FastICA and KICA from Fig. 4.4a and b there is difference between them. This representation shows

**Table 4.1** Few eigenvalues obtained (row wise)

| Value 1 | Value 2 | Value 3 | Value 4 | Value 5 | Value 6 | Value 7 | Value 8 | Value 9 | Value 10 |
|---|---|---|---|---|---|---|---|---|---|
| 9.7170 | 6.7949 | 5.2108 | 4.9132 | 4.1353 | 3.0696 | 2.5667 | 2.3351 | 2.1218 | 1.8188 |
| Value 11 | Value 12 | Value 13 | Value 14 | Value 15 | Value 16 | Value 17 | Value 18 | Value 19 | Value 20 |
| 1.7758 | 1.6045 | 1.4169 | 1.3391 | 1.1514 | 1.0498 | 0.9068 | 0.8708 | 0.8182 | 0.6552 |
| Value 21 | Value 22 | Value 23 | Value 24 | Value 25 | Value 26 | Value 27 | Value 28 | Value 29 | Value 30 |
| 0.4002 | 0.3324 | 0000.0 | 0000.0 | 0000.0 | 00000 | 00000 | 00000 | 00000 | 00000 |

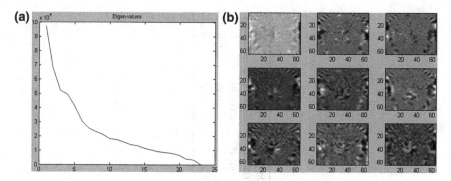

**Fig. 4.3** **a** First 25 eigenvalues. **b** First nine eigenimages

that, how the algorithms provide the different information through ICs and how these ICs are spatially different also it affects on recognition accuracy.

## 4.4 Results with OLOG Edge Detector

In this part, we have used OLOG as edge detector. The orientations we have used in these experimentations are 0°, 45°, 90°, and 135°. The edge of information is extracted using OLOG with four different orientations which are used further for PCA to dimension reduction. Principle component analysis is one of the old and traditional methods of face recognition, but we have done experimentation with PCA and evaluated results on the same database for comparison purpose only. Our interest is to compare the results of PCA, FastICA, and KICA with Indian and Asian face database. The test image and the recognized image from gallery set using PCA algorithm are shown in Fig. 4.5a, b.

The results of PCA algorithm with different orientations of OLOG and L2 distance metrics with variation in pose, illumination, and expressions are shown in Table 4.5. The previous results presented in all sections give idea that L2 distance metric performs better than others. Here we have decided to use only L2 as distance metric. In this experimentation, we have used four sets of principle components like 25, 50, 100, and 200. The results achieved by PCA method are varying from 57 to 80 %. Specifically, the results achieved by combination of PCA + L2 with 90° orientations for pose and expressions variations are almost 80 % when the numbers of principle components used are 25. The comparative analysis of four different orientations of OLOG with L2 metrics is shown in Table 4.5, and same is represented in Fig. 4.6.

After evaluating the results with PCA, next experiment is performed with FastICA algorithm. The face images used here are from Indian and Asian face database. In this study, we have explored feature selection techniques on ICA bases for face recognition. Feature selection techniques are warranted especially

**Table 4.2** First five eigenvector (EV) component values

|     |         |          |          |          |         |         |          |         |         |          |
| --- | ------- | -------- | -------- | -------- | ------- | ------- | -------- | ------- | ------- | -------- |
| EV1 | -0.0013 | 0.002403 | 0.00045  | 0.003921 | -0.0007 | -0.0013 | -0.00356 | 0.00427 | 0.00102 | -0.00065 |
| EV2 | -0.0015 | 0.002172 | 0.00020  | 0.003741 | -0.0008 | -0.0009 | -0.00275 | 0.00352 | 0.00154 | -0.00156 |
| EV3 | -0.0021 | 0.001369 | -0.00029 | 0.002032 | -0.0015 | 0.0001  | -0.02035 | 0.00185 | 0.00250 | -0.00324 |
| EV4 | -0.0038 | 0.001131 | -0.00079 | 0.000397 | -0.0032 | 0.0008  | 0.00257  | 0.00122 | 0.00433 | -0.00573 |
| EV5 | -0.0050 | 0.001285 | -0.00076 | -3.6309  | -0.0044 | 0.0007  | 0.00323  | 0.00165 | 0.00523 | -0.00679 |

**Table 4.3** Values of independent components (IC) obtained by FastICA

| | | | | | | | | | | |
|---|---|---|---|---|---|---|---|---|---|---|
| IC1 | -0.50416 | -0.4766 | -0.28728 | -0.16066 | -0.12886 | -0.17698 | -0.2920 | -0.4040 | -0.42316 | -0.19634 |
| IC2 | -0.028545 | -0.03314 | -0.022572 | -0.01856 | -0.05497 | -0.10108 | -0.1024 | -0.0978 | -0.13976 | -0.1485 |
| IC3 | -0.29101 | -0.27964 | -0.20885 | -0.20932 | -0.21538 | -0.08285 | 0.15951 | 0.3291 | 0.37761 | 0.32676 |
| IC4 | 0.0065052 | 0.021687 | 0.0079818 | -0.04687 | -0.11486 | -0.16756 | -0.2323 | -0.2453 | 0.13984 | 0.96478 |
| IC5 | 0.0095528 | -0.01786 | -0.039831 | -0.03450 | -0.04623 | -0.06869 | -0.0307 | -0.0271 | -0.28303 | -0.56388 |

**Table 4.4** Values of independent components (IC) obtained by KICA

| | | | | | | | | | | |
|---|---|---|---|---|---|---|---|---|---|---|
| IC1 | 0.23119 | 0.13117 | 0.01443 | 0.05683 | 0.16172 | 0.1953 | 0.23119 | 0.13117 | 0.01443 | 0.05683 |
| IC2 | 0.05814 | 0.05929 | 0.06732 | 0.099399 | −0.00815 | −0.4116 | 0.05814 | 0.059295 | 0.067329 | 0.099399 |
| IC3 | −0.3537 | −0.2007 | 0.28723 | 0.60125 | 0.36551 | −0.3047 | −0.3537 | −0.20074 | 0.28723 | 0.60125 |
| IC4 | 0.019927 | −0.1439 | −0.3763 | −0.44945 | −0.2631 | −0.1412 | 0.019927 | −0.14395 | −0.37638 | −0.44945 |
| IC5 | 0.1551 | 0.11661 | −0.0231 | −0.13717 | −0.02423 | 0.22595 | 0.1551 | 0.11661 | −0.02316 | −0.13717 |

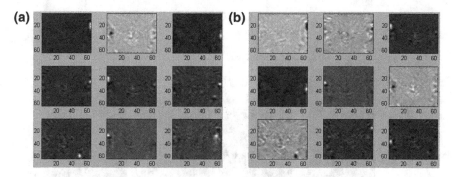

**Fig. 4.4** **a** Few ICs images obtained by FastICA and **b** by KICA

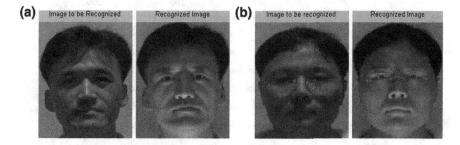

**Fig. 4.5** **a–b** Input and output image of algorithm

**Table 4.5** Result with PCA + OLOG +L2

| No. of principle components | Pose variation with different orientation angle | | | | Illumination variation with different orientations angle | | | | Expression variation with different orientations angle | | | |
|---|---|---|---|---|---|---|---|---|---|---|---|---|
| Orientations | 0 | 45 | 90 | 135 | 0 | 45 | 90 | 135 | 0 | 45 | 90 | 135 |
| 25 | 76 | 76 | 80 | 76 | 80 | 76 | 68 | 72 | 76 | 76 | 80 | 76 |
| 50 | 70 | 68 | 66 | 72 | 72 | 70 | 64 | 64 | 68 | 68 | 72 | 72 |
| 100 | 65 | 64 | 62 | 67 | 66 | 66 | 62 | 61 | 64 | 67 | 67 | 64 |
| 200 | 61 | 59.5 | 61 | 62.5 | 63.5 | 63 | 57.5 | 59 | 61.5 | 62.5 | 62.5 | 61 |

for ICA features, since these are devoid of any importance ranking based on energy content as the PCA components. The test image and the recognized image from gallery set using FastICA algorithm is shown in Fig. 4.7a and b.

With FastICA algorithm, the numbers of independent components used for experimentation are 25, 50, 100, and 200. In this study, we have explored PCA as baseline for FastICA algorithm to reduce dimension of input dataset. In this study, we have used orientations for OLOG are 0°, 45°, 90°, and 135°. The result with

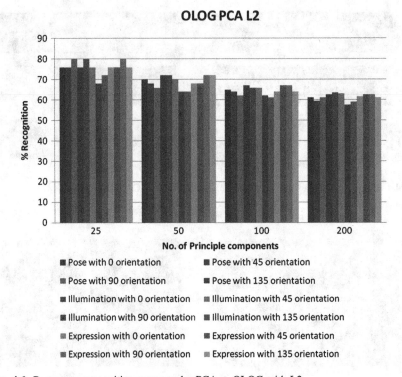

**Fig. 4.6** Percentage recognition accuracy by PCA + OLOG with L2

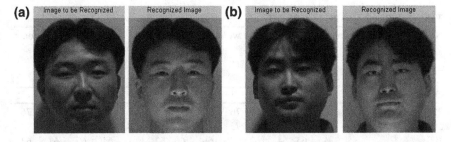

**Fig. 4.7   a–b** Input and output image of algorithm

FastICA varies from 59 to 84 %. The combination of FastICA+L2 gives 80 % results when 25 ICs are used with 0°, 45° orientations for OLOG. The comparative analysis of four different orientations with L2 distance metrics are shown in Table 4.6, and same is represented in Fig. 4.8.

The database known as *Asian face image database* is from Intelligent Multimedia research laboratories having face images of 56 male and 57 female persons with 17 samples each; which consist 4 images of variation of illumination, 8 images with pose variations, and 4 facial expression variation conditions are used

**Table 4.6** Result with FastICA + OLOG +L2

| No. of independent components | Pose variation with different orientation angle | | | | Illumination variation with different orientations angle | | | | Expression variation with different orientations angle | | | |
|---|---|---|---|---|---|---|---|---|---|---|---|---|
| Orientations | 0 | 45 | 90 | 135 | 0 | 45 | 90 | 135 | 0 | 45 | 90 | 135 |
| 25 | 80 | 80 | 80 | 80 | 84 | 84 | 72 | 76 | 80 | 76 | 80 | 76 |
| 50 | 72 | 70 | 68 | 74 | 76 | 76 | 66 | 68 | 76 | 68 | 72 | 70 |
| 100 | 67 | 67 | 65 | 70 | 70 | 72 | 63 | 63 | 70 | 65 | 67 | 65 |
| 200 | 61.5 | 62 | 63 | 63 | 64 | 64.5 | 59 | 60 | 66 | 63 | 61.5 | 62 |

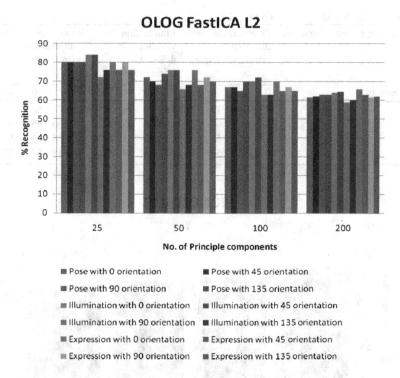

**Fig. 4.8** Percentage recognition accuracy by FastICA + OLOG with L2

for analysis. The resolution of all images used in the algorithm is $128 \times 128$, to reduce computational time. The results with four different orientations of OLOG with L2 classifiers for KICA algorithm is presented for all facial conditions. The test image and the recognized image from gallery set using KICA algorithm is shown in Fig. 4.9a and b.

In this study and analysis with KICA algorithm, the numbers of independent components used for experimentation are 25, 50, 100, and 200. We have explored PCA as baseline for KICA algorithm to reduce dimension and whiten the data

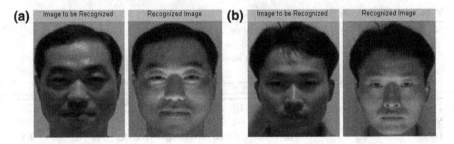

**Fig. 4.9  a–b** Input and output image of algorithm

**Table 4.7**  Result with KICA + OLOG +L2

| No. of independent components | Pose variation with different orientation angle | | | | Illumination variation with different orientations angle | | | | Expression variation with different orientations angle | | | |
|---|---|---|---|---|---|---|---|---|---|---|---|---|
| Orientations | 0 | 45 | 90 | 135 | 0 | 45 | 90 | 135 | 0 | 45 | 90 | 135 |
| 25 | 72 | 68 | 72 | 72 | 76 | 72 | 68 | 68 | 72 | 72 | 68 | 68 |
| 50 | 66 | 64 | 64 | 68 | 68 | 66 | 66 | 64 | 68 | 66 | 66 | 64 |
| 100 | 63 | 62 | 61 | 61 | 63 | 63 | 63 | 62 | 63 | 63 | 63 | 62 |
| 200 | 59 | 57.5 | 59 | 59 | 60 | 59 | 59 | 57.5 | 61 | 59 | 57.5 | 57.5 |

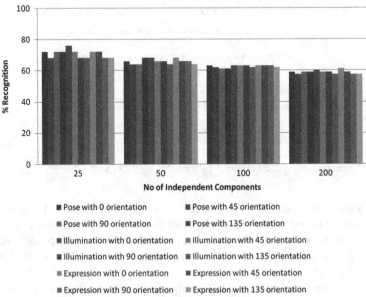

**Fig. 4.10**  Percentage recognition accuracy by KICA + OLOG with L2

matrix. The analysis is performed with four orientations of OLOG with KICA and L2 distance metrics. In our study, the results with KICA for illumination changes variy from 57 to 76 %. The combination of KICA+L2 gives 76 % results when 25 ICs are used with 0° orientations under illumination change. The FastICA algorithm outperforms PCA and KICA algorithms is the indication from analysis presented in this section.

The comparative analysis of four different orientations with L2 distance metrics is presented in Table 4.7, and same is represented in Fig. 4.10.

# Chapter 5
# Multiscale Wavelet-Based Edge Detection for Face Recognition Using ICA

## 5.1 Multiscale Wavelet-Based Edge Detection

This part of the experimentation presents a new approach to edge detection using wavelet transforms. The proposed wavelet-based edge detection algorithm combines the coefficients of wavelet transforms on a series of scales and significantly improves the results of face recognition using wavelet-based edge detection and ICA. The continuous wavelet model constructed by Li [6] not only explains the working mechanism of most classical edge detectors, but also has several significant advantages in practical applications. The scale of the wavelet used in this model can be adjusted to detect edges of different levels of scale. Also, the smoothing function used in the construction of a wavelet reduces the effect of noise. Thus, the smoothing step and edge detection step are combined together to achieve the optimal result. A wavelet transform is computed by convolving the signal with a dilated wavelet. The wavelet transform of $f(x)$ at the scale $s$ and position $x$, computed with respect to the wavelet $\psi^a(x)$, is defined by

$$W_s^a f(x) = f \times \psi_s^a(x) \tag{5.1}$$

The wavelet transform of $f(x)$ with respect to $\psi^b(x)$ is

$$W_s^b f(x) = f \times \psi_s^b(x) \tag{5.2}$$

Therefore, the wavelet transforms $W_s^a f(x)$ and $W_s^b f(x)$ is, respectively, the first and second derivative of the signal smoothed at the scale $s$. The local extrema of $W_s^a f(x)$ [33, 34] thus correspond to the zeros of $W_s^b f(x)$ and to the inflection points of $f_x \theta_s(x)$. The resolution of an image is directly related to the proper scale for edge detection. High resolution and small scale will result in noisy and discontinuous edges; low resolution and large scale will result in undetected edges. The scale is not adjustable with classical edge detectors, but with the wavelet model, we can construct our own edge detectors with proper scales. Because image data is always discrete, the practical scale in images is usually integer. With the cascade algorithm and the wavelet-based edge detection method, we can detect edges of a

K. J. Karande and S. N. Talbar, *Independent Component Analysis of Edge Information for Face Recognition*, SpringerBriefs in Computational Intelligence, DOI: 10.1007/978-81-322-1512-7_5, © The Author(s) 2014

series of integer scales in an image. This can be useful when the image is noisy, or when edges of certain detail or texture are to be neglected. The scale controls the significance of edges to be shown. Edges of higher significance are more likely to be kept by the wavelet transform across scales. Edges of lower significance are more likely to disappear when the scale increases. In this study, we have used scales for wavelet edge detector are 10, 20, 30, and 40.

## 5.2 Continuous Wavelet Transforms

We know that the Fourier transform of a signal $f(t)$ is defined as

$$\hat{f}(w) = \int_{-\infty}^{\infty} f(t)e^{-jwt}\mathrm{d}t. \tag{5.3}$$

since $e^{-jwt}$ is globally supported and not in $L^2$, Fourier analysis is global in nature, which makes it not suitable for detecting local singularities. To obtain the approximate frequency contents of a signal $f(t)$ in the neighborhood of some desired location $t = b$, we can window the function using an appropriate window function $\Phi(t)$ to produce the windowed function $f_b(t) = f(t)\Phi(t - b)$ and then take the Fourier transform of $f_b(t)$. This is called the short-time Fourier transform (STFT). It can be defined at the location $(b, \xi)$ in the time–frequency plane as

$$G_\phi f(b, \xi) = \int_{-\infty}^{+\infty} f(t)\overline{\phi_{b,\xi}(t)}\mathrm{d}t \tag{5.4}$$

where

$$\phi_{b,\xi}(t) = \phi(t - b)e^{i\xi t} \tag{5.5}$$

One can recover the time function $f(t)$ by taking the inverse Fourier transform of $G_\phi f(b, \xi)$, then multiply by $\overline{\phi(t - b)}$ and integrate with respect to $b$:

$$f(t) = \frac{1}{2\pi\|\phi\|^2} \int_{-\infty}^{+\infty} \mathrm{d}\xi e^{i\xi t} \int_{-\infty}^{+\infty} G_\phi f(b, \xi)\overline{\phi(t - b)}\mathrm{d}b \tag{5.6}$$

If we use the Gaussian function as the window function, i.e., $\Phi(t) = \frac{1}{2\Pi\alpha}e^{-\frac{t^2}{4\alpha}}$ ($\alpha > 0$), the transform is called the Gabor transform [6]. If we use a dilated and translated wavelet function as $\Phi_{b,\xi}(t)$, i.e., $\Phi_{b,\xi}(t) = \psi_{b,\xi}(t)$, the transform is a continuous wavelet transform.

## 5.3  Scale of a Wavelet Function

First, we consider some special characteristics of Gaussian filters. A Gaussian function through a Gaussian filter is still a Gaussian function, but with a dilated variance. Let $g_\sigma(x, y)$ be a normalized Gaussian function with variance $\sigma^2$, then

$$
(g_{\sigma 1} \times g_{\sigma 2})(x, y) = \int_{-\infty}^{+\infty} \int_{-\infty}^{+\infty} g_{\sigma 1}(x - h, y - k) g_{\sigma 2}(h, k) dh dk
$$

$$
= \frac{1}{4\pi^2 \sigma_1^2 \sigma_2^2} \int_{-\infty}^{+\infty} \int_{-\infty}^{+\infty} e^{-\frac{(x-h)^2 + (y-k)^2}{2\sigma_1^2}} e^{-\frac{h^2 + k^2}{2\sigma_2^2}} dh dk \tag{5.7}
$$

$$
= \frac{1}{2\pi(\sigma_1^2 + \sigma_2^2)} e^{-\frac{x^2 + y^2}{2(\sigma_1^2 + \sigma_2^2)^2}}
$$

$$
= g\sqrt{\sigma_1^2 + \sigma_2^2}(x - y)
$$

Now, we consider a special class of wavelets, which can be defined as the derivative of a smoothing function, i.e.,

$$
\psi(x, y) = \frac{\partial}{\partial x} \phi(x, y) \tag{5.8}
$$

where $\int_{R^2} \phi(x, y) = 1$.

If the Gaussian is used as the smoothing function, this class of wavelets is called a Gaussian wavelet. From (5.7), we can develop a cascade algorithm to find the wavelet transform of $f(x, y)$ on a series of scales.

$$
\begin{aligned}
W^1 f(ns, x, y) &= \psi_{ns} \times f(x, y) \\
&= \frac{\partial \phi_{ns}}{\partial x} \times f(x, y) \\
&= \frac{\partial \left( \phi_{(n-1)s} \times \phi_s \right)}{\partial x} \times f(x, y) \\
&= \frac{\partial \left( \phi_{(n-1)s} \times f \right)}{\partial x} \times \phi_s(x, y) \\
&= W^1 f((n-1)s, x, y) \times \phi_s(x, y)
\end{aligned} \tag{5.9}
$$

thus, for any integer $n > 1$,

$$
W^1 f(ns, x, y) = \underbrace{\phi_s \times \ldots\ldots\ldots \times \phi_s}_{n-1} \times W^1 f(s, x, y) \tag{5.10}
$$

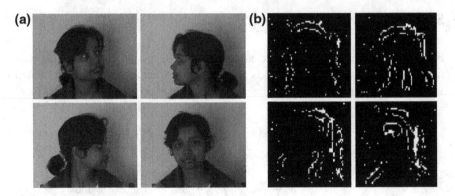

**Fig. 5.1** **a** Input face images. **b** Multiscale wavelet edge detected images scale $= 40$

similarly,

$$W^2 f(ns, x, y) = \underbrace{\phi_s \times \ldots\ldots\ldots \times \phi_s}_{n-1} \times W^2 f(s, x, y) \tag{5.11}$$

The input face images and its multiscale edge detected images are presented in Fig. 5.1a and b, respectively.

## 5.4 Preprocessing by PCA

The preprocessing is briefed in Chap. 1 using PCA. The few sample face images from Indian face database are given in Fig. 5.1a used to find out eigenimages and eigenvalues. Using the covariance matrix, the obtained few eigenvalues are represented in Table 5.1. These values are in descending order and approaches to zero represented row wise in the table. This principle of eigenvalue is used to reduce the dimension of eigenvector matrix, and it becomes the advantage of PCA for dimension reduction. The eigenvalues in the Table 5.1 are represented graphically in Fig. 5.2a.

The eigenvector components values are shown in Table 5.2 and the eigenimages are represented in Fig. 5.2b. Few values of one eigenvector are represented by one row. These eigenimages are nothing but eigenvectors obtained from covariance matrix of input data. The eigenvector matrix is further used by ICA algorithm to extract independent components (ICs).

**Table 5.1** Few eigenvalues obtained (row wise)

| Value 1 | Value 2 | Value 3 | Value 4 | Value 5 | Value 6 | Value 7 | Value 8 | Value 9 | Value 10 |
| --- | --- | --- | --- | --- | --- | --- | --- | --- | --- |
| 246.837 | 177.962 | 163.650 | 154.157 | 140.192 | 132.236 | 126.6784 | 121.0841 | 114.4711 | 108.053 |
| Value 11 | Value 12 | Value 13 | Value 14 | Value 15 | Value 16 | Value 17 | Value 18 | Value 19 | Value 20 |
| 104.2968 | 101.5508 | 97.2483 | 95.6842 | 90.8257 | 90.4646 | 87.8658 | 78.9324 | 76.4243 | 70.4699 |
| Value 21 | Value 22 | Value 23 | Value 24 | Value 25 | Value 26 | Value 27 | Value 28 | Value 29 | Value 30 |
| 67.9986 | 59.5552 | 0000.0 | 0000.0 | 0000.0 | 00000 | 00000 | 00000 | 00000 | 00000 |

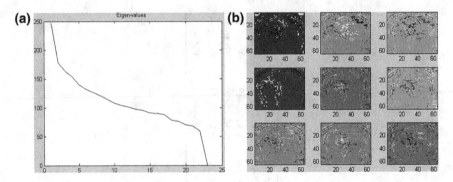

**Fig. 5.2  a** First 25 eigenvalues. **b** First nine eigenimages

## 5.4.1  ICA Algorithms

After preprocessing and getting the eigenvalues and eigenvectors, next step is to calculate the ICs. Few algorithms used to find ICs are fixed point ICA algorithm [28], FastICA algorithm [29], Kernel ICA (KICA) algorithm [30], Joint approximate diagonalization of eigenmatrices (JADE) [17], and Blind source extraction–optimization of kurtosis (BSE-K) algorithm [28]. Here we have used two ICA algorithms i.e., FastICA and KICA algorithms.

The extracted ICs are represented in Table 5.3, where each row represents few values of one independent component by FastICA algorithm. The ICs are represented by images as shown in Fig. 5.3a. The ICs are also extracted by using KICA algorithm and are presented in Table 5.4; the ICs are represented by images as shown in Fig. 5.3b.

If we observe the ICs obtained by FastICA and KICA from Fig. 5.3a and b, there is difference between them.

## 5.5  Results with Multiscale Wavelet Edge Detector

The multiscale waveletbased edge detector is used in this part to extract the edge information. Here the wavelet-based edge detector with multiscale approach is used for edge detector. In this experiment, the scales used are 10, 20, 30, and 40. The validation of the results performed with pose, illumination, and facial expressions variations for face recognition. The extracted edge information is preprocessed by PCA and then ICs are extracted. Principle component analysis is one of the traditional methods of face recognition, but we have studied the usefulness of PCA and evaluated results on the same database for comparison purpose only.

Purpose of this is to compare the results of PCA, FastICA, and KICA with Asian face database. The test image and the recognized image from gallery set

**Table 5.2** First five eigenvector (EV) component values

| EV1 | 6.2617e−1 | −1.213e−1 | 6.9386e−1 | 3.2706e−1 | 3.574e−01 | −6.7712e− | −6.6177 | 9.5729e−1 | −3.7109 | −2.7541e− |
|---|---|---|---|---|---|---|---|---|---|---|
| EV2 | 1.8146 | −2.4831e− | −4.1891e−1 | 1.3077e−1 | 1.3525e−1 | 6.2471e1 | −7.7543 | 3.1259e−1 | −2.1043e1 | 4.4383e−1 |
| EV3 | −1.6639 | 1.5775e−1 | −5.4614 | −6.3855e | 7.3109e− | 7.0866e− | −3.6474e− | 7.7137e− | 1.9868e−1 | −2.2959e− |
| EV4 | 5.8841e−1 | 1.7969e−1 | 3.0627e−1 | −1.6498e− | 1.0645e−1 | 1.2691e−1 | 2.3869e−1 | 4.6448e | −1.6041e− | −3.3264e− |
| EV5 | 9.3083e− | 3.1479e− | 3.0681e− | 7.2954e− | −5.2237e− | 4.0027e− | −6.6052e− | 1.8731e− | 5.0425e− | −2.6115e−0 |

**Table 5.3** Values of independent components (IC) obtained by FastICA

| IC1 | 0 | 0 | 0 | 0 | 0 | −0.23916 | 0 | −0.2391 | −0.36142 | 0 |
|-----|---|---|---|---|---|----------|---|---------|----------|---|
| IC2 | 0 | 0 | 0 | 0 | 0 | −0.23855 | 0 | −0.2385 | −0.41029 | 0 |
| IC3 | 0 | 0 | 0 | 0 | 0 | 0.32075 | 0 | 0.32075 | 0.54033 | 0 |
| IC4 | 0 | 0 | 0 | 0 | 0 | −0.24855 | 0 | −0.2485 | −0.43619 | 0 |
| IC5 | 0 | 0 | 0 | 0 | 0 | 0.2318 | 0 | 0.2318 | 0.43024 | 0 |

**Table 5.4** Values of independent components (IC) obtained by KICA

| IC1 | 0 | 0.31974 | 0 | 0.31974 | −6.2258 | 0 | 0 | 0.31974 | 0 | 0.31974 |
|-----|---|---------|---|---------|---------|---|---|---------|---|---------|
| IC2 | 0 | 0.29648 | 0 | 0.29648 | −0.071855 | 0 | 0 | 0.29648 | 0 | 0.29648 |
| IC3 | 0 | −0.31819 | 0 | −0.31819 | −0.22759 | 0 | 0 | −0.3181 | 0 | −0.3181 |
| IC4 | 0 | −0.35246 | 0 | −0.35246 | −0.090982 | 0 | 0 | −0.3524 | 0 | −0.3524 |
| IC5 | 0 | 0.27487 | 0 | 0.27487 | 0.17666 | 0 | 0 | 0.27487 | 0 | 0.27487 |

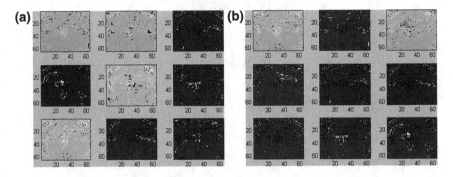

**Fig. 5.3**  **a** Few ICs images obtained by FastICA and **b** by KICA

using PCA algorithm is shown in Fig. 5.4a and b. The results of PCA algorithm with different scales of wavelet edge detector with L2 distance metrics and with variation in pose, illumination, and expressions are shown in Table 5.5. In this experimentation, we have used four sets of principle components like 25, 50, 100, and 200. The results achieved by PCA method are varying from 59 to 84 %. Specifically, the results achieved by combination of PCA + Wavelet + L2 are almost 84 % when the numbers of principle components are used 25 and scale of wavelet is 40. The comparative analysis of four different scale of wavelet with L2 distance metric is shown in Table 5.5, and same is represented in Fig. 5.5.

In first step, we evaluate the results with PCA; in this step experiment is performed with FastICA algorithm. The face images used are from Indian and Asian face database as it has sufficient number of sample images. In this study, we have explored PCA as baseline for ICA for dimension reduction and whitening of data matrix. The test image and the recognized image from gallery set using FastICA algorithm is shown in Fig. 5.6a and b.

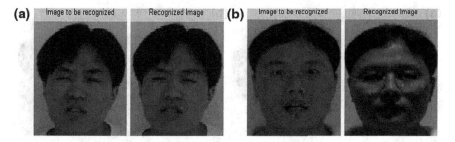

**Fig. 5.4  a–b** Input and output image of algorithm

**Table 5.5**  Result with PCA + wavelet multiscale + L2

| No. of principle components | Pose variation with different scale of wavelet | | | | Illumination variation with different scale of wavelet | | | | Expression variation with different scale of wavelet | | | |
|---|---|---|---|---|---|---|---|---|---|---|---|---|
| Scale of wavelet | 10 | 20 | 30 | 40 | 10 | 20 | 30 | 40 | 10 | 20 | 30 | 40 |
| 25 | 76 | 76 | 80 | 84 | 72 | 80 | 80 | 84 | 76 | 76 | 80 | 80 |
| 50 | 70 | 68 | 70 | 76 | 66 | 74 | 68 | 76 | 70 | 68 | 76 | 72 |
| 100 | 65 | 63 | 67 | 70 | 63 | 70 | 65 | 72 | 65 | 65 | 70 | 67 |
| 200 | 62 | 60 | 62 | 64 | 59 | 63 | 63 | 64.5 | 62 | 63 | 66 | 61.5 |

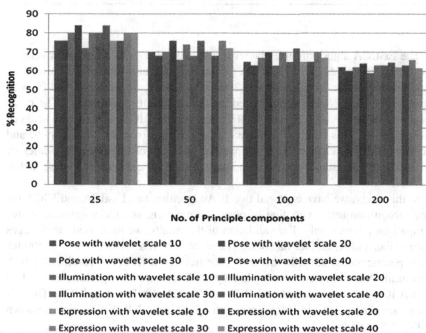

**Fig. 5.5**  Percentage recognition accuracy by PCA + wavelet with L2

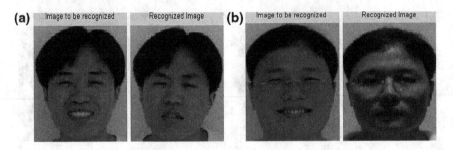

**Fig. 5.6 a–b** Input and output image of algorithm

**Table 5.6** Result with FastICA + wavelet multiscale + L2

| No. of independent components | Pose variation with different scale of wavelet | | | | Illumination variation with different scale of wavelet | | | | Expression variation with different scale of wavelet | | | |
|---|---|---|---|---|---|---|---|---|---|---|---|---|
| Scale of wavelet | 10 | 20 | 30 | 40 | 10 | 20 | 30 | 40 | 10 | 20 | 30 | 40 |
| 25 | 76 | 80 | 84 | 88 | 80 | 84 | 84 | 88 | 84 | 84 | 84 | 88 |
| 50 | 74 | 76 | 78 | 80 | 76 | 78 | 80 | 82 | 80 | 78 | 80 | 82 |
| 100 | 70 | 72 | 72 | 74 | 72 | 70 | 72 | 74 | 74 | 72 | 76 | 78 |
| 200 | 62.5 | 63 | 63.5 | 64 | 64.5 | 64 | 65 | 65.5 | 64 | 63.5 | 64 | 65 |

The FastICA algorithm gives better results with almost all scales of wavelet as shown in Table 5.6. With FastICA algorithm the numbers of ICs used for experimentation are 25, 50, 100, and 200. In this study, we have explored PCA as baseline for FastICA algorithm to reduce dimension of input dataset. The combination of FastICA + L2 gives 88 % results when 25 ICs are used with scale of wavelet is 40. This result is achieved with variation of pose, illumination, and facial expressions change. The comparative analysis of four different scales of wavelet with L2 distance metric is shown in Table 5.6 and same is represented in Fig. 5.7.

In this study, we have explored two ICA algorithm i.e., FastICA and KICA for face recognition under variation of pose, illumination, and facial expressions for comparison purpose only. For validation of the results, we have used face images from Indian and Asian face database. The number of ICs varies from 25 to 200 for the experimentation done. In this study, we have used KPCA as baseline for ICA algorithm. After dimension reduction by KPCA, FastICA algorithm is used to extract ICs. The combination of KPCA + FastICA is referred as KICA. The test image and the recognized image from gallery set using KICA algorithm is shown in Fig. 5.8a and b.

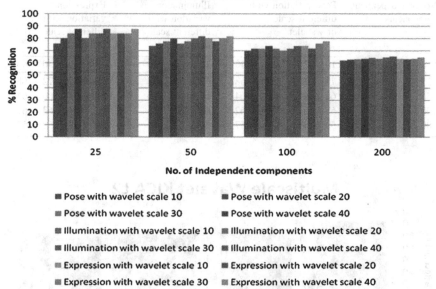

Fig. 5.7 Percentage recognition accuracy by FastICA + wavelet with L2

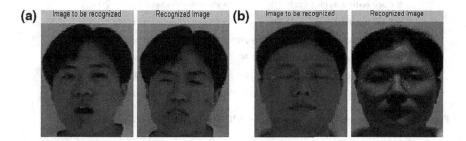

Fig. 5.8 a–b Input and output image of algorithm

The facial expressions used are the different moods of the human being, include happy, sad, disgust, etc. The sufficient number of sample images used from Indian and Asian face database for pose, illumination, and expression changes. In this study and analysis with KICA algorithm, the numbers of ICs used for experimentation are 25, 50, 100, and 200. We have explored PCA as baseline for KICA

**Table 5.7** Result with KICA + wavelet multiscale + L2

| No. of independent components | Pose variation with different scale of wavelet | | | | Illumination variation with different scale of wavelet | | | | Expression variation with different scale of wavelet | | | |
|---|---|---|---|---|---|---|---|---|---|---|---|---|
| Scale of wavelet | 10 | 20 | 30 | 40 | 10 | 20 | 30 | 40 | 10 | 20 | 30 | 40 |
| 25 | 72 | 72 | 68 | 68 | 72 | 72 | 72 | 68 | 72 | 72 | 72 | 68 |
| 50 | 66 | 64 | 64 | 64 | 66 | 64 | 66 | 66 | 68 | 64 | 66 | 64 |
| 100 | 63 | 61 | 63 | 62 | 63 | 61 | 63 | 63 | 63 | 61 | 63 | 62 |
| 200 | 59 | 59 | 57.5 | 57.5 | 59 | 59 | 61 | 59 | 61 | 59 | 59 | 57.5 |

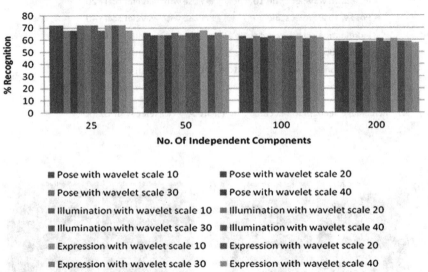

**Fig. 5.9** Percentage recognition accuracy by KICA + wavelet with L2

algorithm to reduce dimension and whiten the data matrix. In our study, the results with KICA for facial expression changes vary from 57 to 72 %. The combination of KICA + L2 gives 72 % results when 25 ICs are used with scales of wavelet are 10, 20, and 30. The comparative analysis of four different scales of wavelet with L2 distance metric is shown in Table 5.7, and same is represented by Fig. 5.9.

# Chapter 6
# Conclusion

In this book, independent component of edge information approach for face recognition is discussed and analyzed. As per the literature survey and our knowledge, there is a hardly reported research work on edge information with ICA for face recognition. This idea of using edge information with ICA algorithms for face recognition is a new approach and provides a different way of analyzing the results for face recognition.

In this study, we have used four edge detection techniques for extracting the edge information. These techniques are Canny, LOG, OLOG, and Multiscale wavelet-based edge detectors. In experimentation, we have extracted edge information from facial images and then preprocessing is performed using PCA. Here, PCA helps for dimension reduction and whitening of matrix. After preprocessing using PCA then independent components are obtained by using FastICA and KICA algorithms. In this experimentation, we extract edge information from facial images and it is used for further processing. After extracting edge information, we lost global information from facial images. Due to loss of global information, the recognition accuracy of face recognition task reduces and it is reflected in the results presented in this research work.

In this study, we have used PCA, FastICA, and KICA algorithms for analyzing the results, and accordingly the comparison is done and presented. In these all experimentations we have used Euclidian distance (L2) as classifiers as it provides better results. If we compare the results of all four different methods of edge detection with ICA algorithm, the maximum recognition accuracy achieved by multiscale wavelet edge detection method is 88 %. The recognition accuracy achieved by Canny and OLOG is also encouraging when number of independent components is 25 and 50. Finally, the overall results of edge information with ICA algorithms are encouraging and it gives the different way of implementation and interpretation of results in the face recognition area.

K. J. Karande and S. N. Talbar, *Independent Component Analysis of Edge Information for Face Recognition*, SpringerBriefs in Computational Intelligence, DOI: 10.1007/978-81-322-1512-7_6, © The Author(s) 2014

# About the Authors

Karande Kailash Jagannath has completed his PhD in Electronics and Telecommunication from SRTM University, Nanded, India; M.Tech. in Electronics and Telecommunication, from Dr. BATU, Lonere, India and B.E. in Electronics and Telecommunication, from Mumbai University. He has total 18 publications in International and National Journals and Conferences. Currently he is working as Principal at SKN Sinhgad College of Engineering, Pandharpur, India. He has offered leadership as coordinator for IGNOU curriculum structure and syllabus design team at National level for Engineering Education.

Sanjay N. Talbar received his B.E and M.E degrees from SGGS Institute of Technology, Nanded, India in 1985 and 1990, respectively. He obtained his PhD (highest degree) from SRTM University, India in 2000. He received the Young Scientist Award by URSI, Italy in 2003. He has collaborative research programme at Cardiff University Wales, UK. Currently he is working as Professor in Electronics and Telecommunication Department of SGGS Institute of Technology Nanded, India. He is a member of many prestigious committees in academic field of India. His research interests are Signal and Image processing and Embedded Systems.

K. J. Karande and S. N. Talbar, *Independent Component Analysis of Edge Information for Face Recognition*, SpringerBriefs in Computational Intelligence, DOI: 10.1007/978-81-322-1512-7, © The Author(s) 2014

# References

1. Gao, Y., Leung, M.K.H.: Face recognition using line edge map. IEEE Trans. Pattern Anal. Mach. Intell. **24**(6), 764–779 (2002)
2. de vel, O., Aeberhard, S.: Line-based face recognition under varying pose. IEEE Trans. Pattern Anal. Mach. Intell. **21**(10), 1081–1088 (1999)
3. Liu, C., Wecher, H.: A shape and texture based enhanced fisher classifier for face recognition. IEEE Trans. Image Process **10**(4), 598–608 (2001)
4. Lee, S.Y., Ham, Y.K., Park, R.H.: Recognition of human front faces using knowledge based feature extraction and neuro fuzzy algorithm. Pattern Recognit **29**(11), 1863–1876 (1996)
5. Burt, P.J., Adelson, E.: The laplacian pyramid as a compact image code. IEEE Trans. Commun. **31**(4), 532–540 (1983)
6. Li, J.: A wavelet approach to edge detection. A thesis presented to The Department of Mathematics and Statistics in partial fulfillment of the requirements for the degree of Master of Science in the subject of Mathematics, Sam Houston State University Huntsville, Texas August (2003)
7. Benedetto, J.J., Frazier, M.W.: Wavelets-Mathematics and Applications. CRC Press, Inc. (1994)
8. Turk, M.A., Pentland, A.P.: Eigenfaces for recognition. J. Cognitive Neurosci. **3**(1), 71–86 (1991)
9. Sirovich, L., Kirby, M.: Low-dimensional procedure for the characterization of human faces. J Opt Soc Am A **4**(3), 519–524 (1987)
10. Bichsel, M., Pentland, A.P.: Human face recognition and the face image set's topology. CVGIP: Image Underst. **59**, 254–261 (1994)
11. Turk, M.: A random walk through eigenspace. IEICE Trans. Inf. Syst. E84-D(12), 1586–1695 (2001)
12. Stevens, K.A.: Surface perception from local analysis of texture and contour. Artificial Intell. Lab., Mass. Instr. Technol., Cambridge Tech. Rep. AI-TR-512 (1980)
13. Moses, Y., Adini, Y., Ullman, S.: Face recognition: The problem of compensating for changes in illumination direction. In: Proceedings of the European Conference on Computer Vision, vol. A, pp. 286–296 (1994)
14. Zhao, W., Chellappa, R., Phillips, P., Rosenfeld, A.: Face recognition: A literature survey. ACM Comput. Surv. **35**(4), 399–458 (2000)
15. Zhange, Y., He, Z.: Blind image restoration using ICA like algorithm. IEEE Int. Conf. Multimedia Expo (2001)
16. Scholkopf, B., Smola, A., Muller, K.R.: Nonlinear component analysis as a kernel eigenvalue problem. Neural Comput. **10**(5), 1299–1319 (1998)
17. Turk, M.A., Pentaland, A.P.: Face recognition using eigenfaces. IEEE conf. Comput. Vis. Pattern Recognition, pp. 586–591 (1991)

K. J. Karande and S. N. Talbar, *Independent Component Analysis of Edge Information for Face Recognition*, SpringerBriefs in Computational Intelligence, DOI: 10.1007/978-81-322-1512-7, © The Author(s) 2014

18. Ekenel, H.K., Sankur, B.: Feature selection in the independent component subspace for face recognition. Pattern Recogn. Lett. (Elsevier) **25**, 1377–1388 (2004)
19. Akaho, S.: Conditionally independent component analysis for supervised feature extraction. Neurocomput **49**, 139–150 (2002). Elsevier
20. Kawakatsu, M.: Application of ICA to MEG noise reduction. In: 4th International Symposium on Independent Component Analysis and Blind Signal Separation (ICA 2003), Nara, Japan, April 2003
21. La Cascia, M., Sclaroff, S.: Fast, reliable head tracking under varying illumination. IEEE Trans. Pattern Anal. Mach. Intell. **22**(4), (2000)
22. Qiu, B., Prinet, V., Perrier, E., Monga, O.: Multi-block PCA method for image change detection. In: 12th IEEE International Conference on Image Analysis and Processing (ICIAP), 2003
23. Barret, M., Narozny, M.: Application of ICA to lossless image coding. In: 4th International Symposium on Independent Component Analysis and Blind Signal Separation (ICA 2003), Nara, Japan, April 2003
24. Umeyama, S.: Blind deconvolution of images using gabor filters and independent component analysis. In: 4th International Symposium on Independent Component Analysis and Blind Signal Separation (ICA 2003), Nara, Japan, April 2003
25. Gonzalez, R.C., Woods, R.E.: Digital image processing, 2nd edn. Pearson Education, New Jersey (2003)
26. Turk, M.A., Pentaland, A.P.: Face recognition using eigenfaces. In: IEEE conference on Computer Vision and Pattern Recognition, pp. 586–591 (1991)
27. Phillips, P.J., et al.: Overview of the face recognition grand challenge. In: IEEE Computer Society Conference on Computer Vision and Pattern Recognition CVPR, 2005
28. Hyvarinen, A., Karhunen, J., Oja, E.: Independent component analysis. Wiley Interscience Publication, Wiley, inc, New York (2002)
29. Hyvärinen, A., Oja, E.: Independent component analysis: Algorithms and applications. Neural Networks Research Centre Helsinki University of Technology P.O. Box 5400, FIN-02015 HUT, Finland, Neural Networks, **13**(4–5), 411–430 (2000)
30. Yang, J., Gao, X., Zhang, D., Yang, J.: Kernel ICA: An alternative formulation and its application to face recognition. Pattern Recogn. J. Pattern Recogn. Soc. **38**, 1784–1787 (2005)
31. Turk, M., Pentland, A.: Eigenfaces for recognition. J. Cognitive Neurosci. **3**(1), 71–86 (1991)
32. Greenspan, H., Goodman, R., Chellapa, R., Anderson, C.H.: Learning texture discrimination rules in a multi resolution system. In: IEEE Trans. Pattern Recogn. Mach. Intell. **16**(9), 894–901 (1994)
33. Mallat, S., Hwang, W.L.: Singularity detection and processing with wavelets. IEEE Trans. Inform. Theory **38**(2), 617–64 (1992)
34. Tang, Y.Y., Yang, L., Liu, J.: Characterization of dirac-structure edges with wavelet transform. IEEE Trans. Sys. Man Cybern. B Cybern **30**(1), 93–109 (2000)
35. Karande, K.J., Talbar, S.N.: Face recognition under variation of pose and illumination using independent component analysis. ICGST's J. Graphics Vision Image Process. (GVIP) ICGST-GVIP **8**(IV), 1–6 (2008)
36. Karande, K.J., Talbar, S.N.: Independent component analysis of edge information for face recognition. Int. J. Image Process. (IJIP), J. Comput. Sci. **3**(3), 120–131 (2009)
37. Karande, K.J., Talbar, S.N., Inamdar, S.S.: Face recognition using oriented laplacian of gaussian (OLOG) and independent component analysis (ICA). In: The 2nd IEEE International Conference on Digital Information and Communication Technology and its Applications (DICTAP 2012), at University of the Thai Chamber of Commerce (UTCC), Bangkok, Thailand, pp. 99–103, 16–18 May 2012
38. Karande, K.J.: Modified modular independent component analysis (ICA) approach for face recognition. In: International Conference on Information Technology, Electronics and Communications (ICITES) at Hyderabad, pp. 114–119, 14th and 15th July 2012

39. Karande, K.J., Talbar, S.N.: Simplified and Modified Approach for Face Recognition using PCA. In: IEEE International Conference on Information and Communication Technology in Electrical Sciences ICTES 2007 at Chennai by IET UK, 20–22 Dec 2007
40. Karande, K.J.: Multiscale wavelet based edge detection and Independent Component Analysis (ICA) for face recognition. In: IEEE 2012 International Conference on Communication, Information and Computing Technology (ICCICT) at SPIT Mumbai, India, Oct 19–20
41. Karande, K.J.: Localized spatiotemporal modular ICA for face recognition. Accepted for IEEE Symposium Series on Computational Intelligence 2013 (IEEE SSCI 2013), Singapore on 15th to 19th April 2013